AF897827

Computational Social Sciences

Computational Social Sciences

A series of authored and edited monographs that utilize quantitative and computational methods to model, analyze and interpret large-scale social phenomena. Titles within the series contain methods and practices that test and develop theories of complex social processes through bottom-up modeling of social interactions. Of particular interest is the study of the co-evolution of modern communication technology and social behavior and norms, in connection with emerging issues such as trust, risk, security and privacy in novel socio-technical environments.

Computational Social Sciences is explicitly transdisciplinary: quantitative methods from fields such as dynamical systems, artificial intelligence, network theory, agent-based modeling, and statistical mechanics are invoked and combined with state-of-theart mining and analysis of large data sets to help us understand social agents, their interactions on and offline, and the effect of these interactions at the macro level. Topics include, but are not limited to social networks and media, dynamics of opinions, cultures and conflicts, socio-technical co-evolution and social psychology. Computational Social Sciences will also publish monographs and selected edited contributions from specialized conferences and workshops specifically aimed at communicating new findings to a large transdisciplinary audience. A fundamental goal of the series is to provide a single forum within which commonalities and differences in the workings of this field may be discerned, hence leading to deeper insight and understanding.

Series Editors:

Elisa Bertino
Purdue University, West Lafayette, IN, USA

Claudio Cioffi-Revilla
George Mason University, Fairfax, VA, USA

Jacob Foster
University of California, Los Angeles, CA, USA

Nigel Gilbert
University of Surrey, Guildford, Surrey, UK

Jennifer Golbeck
University of Maryland, College Park, MD, USA

Bruno Gonçalves
New York University, New York, NY, USA

James A. Kitts
University of Massachusetts, Amherst, MA, USA

Larry S. Liebovitch
Queens College, City University of New York, New York, NY, USA

Sorin A. Matei
Purdue University, West Lafayette, IN, USA

Anton Nijholt
University of Twente, Enschede, The Netherlands

Andrzej Nowak
University of Warsaw, Warsaw, Poland

Robert Savit
University of Michigan, Ann Arbor, MI, USA

Flaminio Squazzoni
University of Brescia, Brescia, Brescia, Italy

Alessandro Vinciarelli
University of Glasgow, Glasgow, Scotland, UK

More information about this series at http://www.springer.com/series/11784

Federico Cecconi • Marco Campennì
Editors

Information and Communication Technologies (ICT) in Economic Modeling

Springer

Editors
Federico Cecconi
LABSS
ISTC-CNR
ROME, Italy

Marco Campennì
Biosciences
University of Exeter
Penryn, Cornwall, UK

ISSN 2509-9574 ISSN 2509-9582 (electronic)
Computational Social Sciences
ISBN 978-3-030-22604-6 ISBN 978-3-030-22605-3 (eBook)
https://doi.org/10.1007/978-3-030-22605-3

© Springer Nature Switzerland AG 2019
This work is subject to copyright. All rights are reserved by the Publisher, whether the whole or part of the material is concerned, specifically the rights of translation, reprinting, reuse of illustrations, recitation, broadcasting, reproduction on microfilms or in any other physical way, and transmission or information storage and retrieval, electronic adaptation, computer software, or by similar or dissimilar methodology now known or hereafter developed.
The use of general descriptive names, registered names, trademarks, service marks, etc. in this publication does not imply, even in the absence of a specific statement, that such names are exempt from the relevant protective laws and regulations and therefore free for general use.
The publisher, the authors, and the editors are safe to assume that the advice and information in this book are believed to be true and accurate at the date of publication. Neither the publisher nor the authors or the editors give a warranty, express or implied, with respect to the material contained herein or for any errors or omissions that may have been made. The publisher remains neutral with regard to jurisdictional claims in published maps and institutional affiliations.

This Springer imprint is published by the registered company Springer Nature Switzerland AG
The registered company address is: Gewerbestrasse 11, 6330 Cham, Switzerland

Contents

Part I
Theory

Chapter 1
Agent-Based Computational Economics and Industrial Organization Theory

Claudia Nardone

Abstract Agent-based computational economics (ACE) is "the computational study of economic processes modeled as dynamic systems of interacting agents." This new perspective offered by agent-based approach makes it suitable for building models in industrial organization (IO), whose scope is the study of the strategic behavior of firms and their direct interactions. Better understanding of industries' dynamics is useful in order to analyze firms' contribution to economic welfare and improve government policy in relation to these industries.

Keywords Agent-based computational economics · Industrial organization theory · Bounded rationality · Complexity · Strategic behavior of firms

Introduction

According to the official definition given by Leigh Tesfatsion (2006), agent-based computational economics (ACE) is "the computational study of economic processes modeled as dynamic systems of interacting agents."

This definition leads straight to the "core business" of this approach, which makes it different from the other ones: economies are considered as complex, adaptive, dynamic systems, where large numbers of heterogeneous agents interact through prescribed rules, according to their current situation and the state of the world around them. Thus, rather than relying on the assumption that the economy will move toward an equilibrium state, often predetermined, ACE aims to build models based on more realistic assumptions. In this way, it is possible to observe if and how an equilibrium state will be reached, and how macro-outcomes will come out, not as a consequence of a typical isolated individual

C. Nardone (✉)
CEIS – Centre for Economic and International Studies, Faculty of Economics – University of Rome "Tor Vergata", Rome, Italy
e-mail: claudia.nardone@uniroma2.it

© Springer Nature Switzerland AG 2019
F. Cecconi, M. Campennì (eds.), *Information and Communication Technologies (ICT) in Economic Modeling*, Computational Social Sciences,
https://doi.org/10.1007/978-3-030-22605-3_1

behavior, but from direct endogenous interactions among heterogeneous and autonomous agents.

This new perspective offered by agent-based approach makes it suitable for building models in industrial organization (IO), whose scope is the study of the strategic behavior of firms and their direct interactions. Better understanding of industries' dynamics is useful in order to analyze firms' contribution to economic welfare and improve government policy in relation to these industries.

In this chapter main features of agent-based computational economics (ACE) will be presented, and some active research areas in this context will be shown, in order to illustrate the potential usefulness of the ACE methodology. Then, we will discuss the main ingredients that tend to characterize economic AB models and how they can be applied to IO issues.

Agent-Based Computational Approach

Traditional quantitative economic models are often characterized by fixed decision rules, common knowledge assumptions, market equilibrium constraints, and other "external" assumptions. Direct interactions among economic agents typically play no role or appear in the form of highly stylized game interactions. Even when models are supported by microfoundations, they refer to a representative agent that is considered rational and makes decisions according to an optimizing process. It seems that economic agents in these models have little room to breathe.

In recent years, however, substantial advances in modeling tools have been made, and economists can now quantitatively model a wide variety of complex phenomena associated with decentralized market economies, such as inductive learning, imperfect competition, endogenous trade network formation, etc. One branch of this new work has come to be known as agent-based computational economics (ACE), i.e., the computational study of economies modeled as evolving systems of autonomous interacting agents. ACE researchers rely on computational frameworks to study the evolution of decentralized market economies under controlled experimental conditions.

Any economy should be described as a complex, adaptive, and dynamic system (Arthur et al. 1997): complexity arises because of the dispersed and nonlinear interactions of a large number of heterogeneous autonomous agents – one of the objectives of ACE is to examine how the macro-outcomes that we can naturally observe arise starting from not examining the behavior of a typical individual in isolation. Global properties emerge instead from the market and non-market interactions of people without them being part of their intentions (Holland and Miller 1991).

In economics, the complexity approach can boast a long tradition, made of many different economists and their theories, starting from the early influence of Keynes and von Hayek and continuing to Schelling and Simon. See for example Keynes (1956), Von Hayek (1937), Schelling (1978). The shift of perspective brought in by full comprehension of their lesson has two implications for economic theory. The first deals with the assumption of rationality used to model human decision-making. By their very nature, optimization techniques guarantee the correspondence of substantive and procedural rationality if and only if all the consequences of alternative actions can be consistently conceived in advance, at least in a probabilistic sense. For complex systems, this possibility is generally ruled out, as interactive population dynamics gives rise to uncertainty that could not be reduced to risk or to a set of probabilities.

Non-cooperative game theory (Shubik 1975) tried to find solutions, but in games with players that are heterogeneous as regards their strategy and their information sets, full adherence to strategic behavior modeling returns computationally complex problems. Solution time for them (measured as the number of simple computational steps required to solve it) increases exponentially in the problem size. As the number of players increases, the size of the problem is too large to complete a search for an optimal solution within a feasible time horizon.

In large interactive systems, individual decision processes become unavoidably adaptive, which is adjusted in the light of realized results, and the search for actions aimed at increasing individual performance stops as soon as a satisfying solution has been found (Simon 1987). Adaptation is backward-looking, sequential, and path-dependent. Desired prices, quantities, inventories, and even the identity of whom we would like to trade are updated according to "error-correction" procedures. Expectations on the future course of events and results are clearly an important part of the decision-making process, but foresights are taken over finite horizons and are modified sequentially in the light of realized outcomes. In complex economies, the key driver of evolution is not optimization but selection. Therefore, in modeling economics from a complex perspective, bounded rationality should be the rule.

The second implication of the complexity approach deals with the common practice of closing models through the exogenous imposition of a general equilibrium solution by means of some fixed-point theorems. Market outcomes must be derived from the parallel computations made by a large number of interacting, heterogeneous, adaptive individuals, instead of being deduced as a fixed-point solution to a system of differential equations. The process of removal of externally imposed coordination devices induces a shift from a top-down perspective toward a bottom-up approach (Delli Gatti et al. 2011). Sub-disciplines of computer science like distributed artificial agent intelligence and multi-agent systems are natural fields to look at. Agent-based computational economics represents a promising tool for advancements along the research program sketched so far.

The ABC approach allows us to build models with a large number of heterogeneous agents, where the resulting aggregate dynamics is not known a priori and outcomes are not immediately deducible from individual behavior.

As in a laboratory experiment, the ACE modeler starts by constructing an economy comprising an initial population of agents (Tesfatsion 2003). These agents can include both economic agents (e.g., consumers, producers, intermediaries, etc.) and agents representing various other social and environmental phenomena (e.g., government agencies, land areas, weather, etc.). The ACE modeler specifies the initial conditions and the attributes of any agent, such as type characteristics, internalized behavioral norms, internal modes of behavior (including modes of communication and learning), and internally stored information about itself and other agents. The economy then evolves over as its constituent agents repeatedly interact with each other and learn from these interactions, without further intervention from the modeler. All events that subsequently occur must arise from the historical timeline of agent-agent interactions.

Main Features

What follows is a sketch of main features that an agent-based model must have, to be defined so. We follow Fagiolo and Roventini (2012, 2016) who describe the main ingredients that usually characterize economic AB models.

1. A bottom-up perspective. As we said, the outcome of the model and the aggregate properties must be derived from direct interactions between agents, without any external or "from above" intervention. This contrasts with the top-down nature of traditional neoclassical models, where the bottom level typically comprises a representative individual, which is constrained by strong consistency requirements associated with equilibrium and hyper-rationality.
2. Heterogeneity. Agents are (or might be) heterogeneous in almost all their characteristics, both attributes and behavioral norms, i.e., how they interact with other agents and the way they learn from their past and from what happens around them.
3. Direct endogenous interactions. Agents interact directly, according to some behavioral norms initially defined, which can evolve through time. The decisions undertaken today by an agent directly depend, through adaptive expectations, on the past choices made by itself and the other agents in the population.
4. Bounded rationality. Generally, in agent-based models, the environment in which agents live is too complex for hyper-rationality to be a viable simplifying assumption, so agents are assumed to behave as bounded rational entities with adaptive expectations. Bounded rationality arises both because information is private and limited and because agents are endowed with a finite computing capacity.
5. Learning process. In AB models, agents are characterized by the ability to collect available information about the current and past state of a subset of other agents and about the state of the whole economy. They put this knowledge into routines and algorithmic behavioral rules. This is the so-called process of "learning," through which agents dynamically update their own state to better perform and achieve their goals. Behavioral rules are not necessarily optimizing in a narrow sense, because, by their very nature, optimization techniques guarantee the correspondence of substantive and procedural rationality if and only if all the consequences of alternative actions can be consistently conceived in advance, at least in a probabilistic sense. For complex systems, this possibility is generally ruled out, as interactive population dynamics implies uncertainty that could not

be reduced to risk or to a set of probabilities. In large interactive systems, individual decision processes become unavoidably adaptive, i.e., adjusted in the light of realized results.

6. Nonlinearity. The interactions that occur in AB models are inherently nonlinear. Additionally, nonlinear feedback loops exist between micro- and macro-levels.
7. The evolving complex system (ECS) approach. Agents live in a complex system that evolves through time. During the repeated interactions among agents, aggregate properties emerge out and can change the environment itself, as well as the way the agents interact.
8. "True" dynamics. Partly because of adaptive expectations (i.e., agents observe the past and form expectations about the future based on the past), AB models are characterized by nonreversible dynamics: the state of the system evolves in a path-dependent manner.

Some Literature References

The last two decades have seen rapid growth of agent-based modeling in economics. Here some of the active research areas that use agent-based computational paradigm are presented.

Macroeconomic Policy in ABMs

ABMs configure themselves as a very powerful device to address policy questions, because of their realistic, flexible, and modular frameworks. Furthermore, an increasing number of leading economists have claimed that the 2008 "economic crisis is a crisis for economic theory" (e.g., Kirman 2010, 2016; Colander et al. 2009; Krugman 2009; Farmer and Foley 2009; Stiglitz 2011, 2015; Kay 2011; Dosi 2012; Romer 2016). Their view is that the predominant theoretical framework, the so-called new neoclassical synthesis (Goodfriend and King 1997), grounded on dynamic stochastic general equilibrium (DSGE) models, isn't able to replicate existing reality and so to explain what actually happens in the economy. These models suffer from a series of dramatic problems and difficulties concerning their inner logic consistency and the way they are taken to the data. In particular, basic assumptions of mainstream DSGE models, which are rational expectations, representative agents, perfect markets, etc., prevent the understanding of basic phenomena underlying the current economic crisis and, more generally, macroeconomic dynamics. For all these reasons, the number of agent-based models dealing with macroeconomic policy issues is increasing fast over time. As the title of a well-known *Nature* article reads, "the economy needs agent-based modelling" (Farmer and Foley 2009).

Dosi et al. (2010, 2017) try to jointly study the short- and long-run impact of fiscal policies, developing an agent-based model that links Keynesian theories of demand generation and Schumpeterian theories of technology-fueled economic growth. Their model is populated by heterogeneous capital-good firms, consumption good firms, consumers/workers, banks, Central Bank, and a public sector. Each agent plays the same role it plays in the real world, so capital-good firms perform R&D and sell heterogeneous machine tools to consumption-good firms and consumers supply labor to firms and fully consume the income they receive. Banks provide credit to consumption-good firms to finance their production and investment decisions. The Central Bank fixes the short-run interest rate and the government levies taxes, and it provides unemployment benefits. The model is able to endogenously generate growth and replicate an ensemble of stylized facts concerning both macroeconomic dynamics (e.g., cross-correlations, relative volatilities, output distributions) and microeconomic ones (firm size distributions, firm productivity dynamics, firm investment patterns). After having been empirically validated according to the output generated, the model is employed to study the impact of fiscal policies (i.e., tax rate and unemployment benefits) on average GDP growth rate, output volatility, and unemployment rate. The authors find that Keynesian fiscal policies are a necessary condition for economic growth and they can be successfully employed to dampen economic fluctuations.

Another paper that moves from a discussion of the challenges posed by the crisis to standard macroeconomics is Caiani et al. (2016). The authors argue that a coherent and exhaustive representation of the inter-linkages between the real and financial sides of the economy should be a pivotal feature of every macroeconomic model and propose a macroeconomic framework based on the combination of the agent-based and stock flow consistent approaches. They develop a fully decentralized AB-SFC model and thoroughly validate it in order to check whether the model is a good candidate for policy analysis applications. Results suggest that the properties of the model match many empirical regularities, ranking among the best performers in the related literature, and that these properties are robust across different parameterizations. Furthermore, the authors state that their work has also a methodological purpose because they try to provide a set of rules and tools to build, calibrate, validate, and display AB-SFC models.

Financial Markets

Financial markets have become one of the most active research areas for ACE modelers. As LeBaron (2006) shows, in an overview of the first studies in this area, financial markets are particularly appealing applications for agent-based methods for several reasons. They are large well-organized markets for trading securities which can be easily compared. Currently, the established theoretical structure of market efficiency and rational expectations is being questioned. There is a long list of empirical features that traditional approaches have not been able to match. Agent-based approaches provide an intriguing possibility for solving some of these puzzles. Finally, financial markets are rich in data sets that can be used for testing and calibrating agent-based models. High-quality data are available at many frequencies and in many different forms.

Models in the realm of agent-based computational finance view financial markets as interacting groups of learning, boundedly rational agents. In these worlds, bounded rationality is driven by the complexity of the state space more than the perceived limitations of individual agents. In agent-based financial markets, dynamic heterogeneity is critical. This heterogeneity is represented by a distribution of agents, or wealth, across either a fixed or changing set of strategies. In principle, optimizing agents would respond optimally to this distribution of other agent strategies, but in general, this state space is far too complicated to begin to calculate an optimal strategy, forcing some form of bounded rationality on both agents and the modeler.

Arthur et al. (1996) developed the highly influential Santa Fe artificial stock market, proposing a dynamic theory of asset pricing based on heterogeneous stock market traders who continually adapt their expectations individually and inductively. According to the authors, "agents forecasts create the world agents are trying to forecast." This means that agents can only treat their expectations as hypotheses: they act inductively, generating individual expectational models that they constantly introduce, test, act upon, and discard. The market becomes driven by expectations that adapt endogenously to the ecology these expectations cocreate.

A more recent survey of agent-based modeling for finance is Cristelli et al. (2011) which discuss, in a unified framework, a number of influential agent-based models for finance with the objective of identifying possible lines of convergence. Models are compared both in terms of their realism and their tractability. A broader perspective can be found in Chen (2012) which gives a historical overview of how agent-based computational economics has developed looking at four origins: the market, cellular automata, tournaments (or game theoretic), and experiments. In thinking about financial markets, the first is of most obvious relevance, but work stemming from all four approaches has played a role in the agent-based modeling of financial markets. The market, understood as a decentralized process, has been a key motivation for agent-based work; Chen argues that the rise of agent-based computational economics can be understood as an attempt to bring the ideas of many and complex heterogeneous agents back into economic consideration.

Electricity Markets

Another very active research area which uses agent-based computational approach to model the dynamics of a single industry is ACE literature on electricity markets. In the last decade, large efforts have been dedicated to developing computational approaches to model deregulated electricity markets, and ACE has become a reference paradigm for researchers working on these topics.

Some researchers have applied agent-based models for examining electricity consumer behavior at the retail level, for example, Hämäläinen et al. (2000), Roop and Fathelrahman (2003), Yu et al. (2004), and Müller et al. (2007). Others study distributed generation models, for example, Newman et al. (2001), Rumley et al. (2008), and Kok et al. (2008).

The topic that has been the major strand of research in this field is wholesale electricity market models. By its nature, ACE is able to take into account several aspects of the procurement process, i.e., all economic events occurring among customers and suppliers during actual negotiations and trading processes. In wholesale electricity markets, mainly characterized by a centralized market mechanism such as the double auction, these aspects are crucial to study the market performance and efficiency but also to compare different market mechanisms. ACE researchers place great confidence in providing useful and complementary insights into the market functioning by a "more realistic" modeling approach. A critical survey of agent-based wholesale electricity market models is Guerci, Rastegar, and Cincotti (2010).

ABM and Industrial Organization Theory

Strategic interactions of economic agents (such as individuals, firms, institutions), i.e., taking into account other agents' actions into their own decision-making processes, are the basis of industrial organization (IO) theory. As in IO theory, agents in ACE models can be represented as interactive goal-directed entities, strategically aware of both competitive and cooperative possibilities with other agents. Moreover, ACE approach offers the key advantage of being able to define heterogeneous agents with a heterogeneous set of properties and behaviors and, as in the behavioral game theory, with the ability to learn, by changing their behavior (response functions) based on previous experience, and thus evolve. In this sense, agent-based tools facilitate to include real-world aspects, such as asymmetric information, imperfect competition, and externalities, which are crucial in IO theory, but often difficult to manage.

Another advantage of the agent-based approach, as Delli Gatti et al. (2011) show, is that modeling can proceed even when equilibria are computational intractable or non-existent: agent-based simulations can handle a far wider range of nonlinear behavior than conventional equilibrium models. Furthermore, there is the possibility to acquire a better understanding of economic processes, local interactions, and out-of-equilibrium dynamics (Arthur, 2006). So, it can be a useful tool where the analytical framework isn't able to find a solution.

Although there are similarities, there is a lack of integration between agent-based approach and the industrial organization literature. There are still few works that use ACE approach to model different market settings or to study market equilibrium in different competition conditions.

An interesting work, which represents an attempt to combine ACE and classic models of IO theory, is Barr and Saraceno (2005). They apply agent-based modeling to Cournot competition, in order to investigate the effects of both environmental and organizational factors on repeated Cournot game outcome. In this model, firms with different organizational structures compete à la Cournot. Each firm is an information processing network, able to learn a whole data set of environmental variables and make its optimal output decision based on these signals, which then influence the demand function. Firms are modeled as a type of artificial neural network (ANN), to make explicit organizational structure and hence to include it in a model of firm competition. Then, they investigate the relationship between optimal firm structure, defined as the most proficient in learning the environmental characteristics, and the complexity of the environment in which quantity competition takes place. Results show that firms modeled as neural networks converge to the Nash equilibrium of a Cournot game: over time, firms learn to perform the mapping between environmental characteristics and optimal quantity decisions. The conclusion is that the optimal firm size is increasing in the complexity of the environment itself and that in more complex environments the necessary time to learn shaping demand factors is longer.

Other attempts to describe theoretical microeconomic models through agent-based approach are represented by Chang (2011), who analyzes entry and exit in an industrial market characterized by turbulent technological processes and by quantity competition, examining how industry-specific factors give rise to across-industries differences in turnover. Rixen and Weigand (2014) study the diffusion of smart meters, considering suppliers who act strategically according to Cournot competition and testing the effects on speed and level of smart meter adoption, if different policies are introduced, such as market liberalization, information policies, and monetary grants. However, all these studies rely on the equilibrium equations of the theoretical models, so the simulated markets are constrained by the theoretical assumptions. A recent interesting work of Sanchez-Cartas (2018) develops an agent-based algorithm based on Game Theory that allows simulating the pricing in different markets, showing that the algorithm is capable of simulating the optimal pricing of those markets. In this way, he tries to overcome difficulties due to the strategic nature of prices, which limits the development of agent-based models with endogenous price competition and helps to establish a link between the industrial organization literature and agent-based modeling. Other studies that exploit agent based approach to model industrial organization dynamics are: Diao et al. (2011), Zhang and Brorsen (2011), van Leeuwen and Lijesen (2016).

In Chap. 5 an agent-based model is developed to mimic trading between firms in a supply chain. Agents are firms who lay on different levels of the chain and are engaged in trading. At each level, firms buy the input from firms at the previous level and sell on the half-processed good to firms at the subsequent level. We are interested in what happens to prices when firms with capacity constraints compete both in price and quantity at the same time. We then introduce, at a certain production stage, a "cartel": some or all firms collude and set a price above the competitive level. In this way, we are able to quantify the pass-on rate, i.e., the proportion of the illegal price increase that cartel direct purchasers, in turn, translate into an increase in their own final price. The extent of the cost translation into prices substantially varies from one setting to another, because it strictly depends on a huge set of different factors, such as market structure, the degree of competition, buyer power, dynamic changes in competition, different prices strategies, etc. To quantify the true pass-on rate, it is thus necessary to take into account all these aspects together. Here, we consider different numbers of firms involved in the illicit agreement and see how the pass-on rate changes in different scenarios.

In this model, we therefore try to solve some computational and behavioral problems in production chain pricing, not easily solvable within analytical frameworks, such as rationing processes, combined with the "minimum price" rule, and best responses to rationing processes.

Conclusions

Agent-based computational economics represents an alternative paradigm or, at least, a complement for analytical modeling approaches. It is characterized by three main tenets: (i) there is a multitude of objects that interact with each other and with the environment; (ii) objects are autonomous (hence they are called "agents"); no central or "top-down" control over their behavior is admitted; and (iii) the outcome of their interaction is computed numerically. Starting from initial conditions, specified by the modeler, the computational economy evolves over time as its constituent agents repeatedly interact with each other and learn from these interactions. ACE is therefore a bottom-up culture-dish approach to the study of economic systems.

Thanks to the possibility to introduce more realistic assumptions but also after the "crisis" that traditional economics has passed in the last years, the agent-based approach has seen rapid growth in some research areas such as macroeconomic policy, financial markets, and electricity markets. However, this approach isn't still as widespread as it deserves. Despite the widespread interest in ABM approaches, it remains at the fringe of mainstream economics. As Rand and Rust (2011) state:

> Despite the power of ABM, widespread acceptance and publication of this method in the highest-level journals has been slow. This is due in large part to the lack of commonly accepted standards of how to use ABM rigorously.

This problem is not new, but although some advances are taking place, there is plenty of room for improvement.

References

Arthur, W. B. (2006). Out-of-equilibrium economics and agent-based modeling. In *Handbook of computational economics*. Elsevier, Amsterdam, North Holland (Vol. 2, pp. 1551–1564).

Arthur, W. B., Holland, J. H., LeBaron, B., Palmer, R., &Taylor, P. (1996). Asset pricing under endogenous expectation in an artificial stock market. Santa Fe Institute, Working Paper No. 96-12-093.

Arthur, W. B., Kollman, K., Miller, J., Page, S., Durlauf, S. N., & Lane, D. A. (1997). Computational political economy. In *The economy as an evolving complex system II* (Vol. 17, pp. 461–490).

Barr, J., & Saraceno, F. (2005). Cournot competition, organization and learning. *Journal of Economic Dynamics and Control, 29*(1–2), 277–295.

Caiani, A., Godin, A., Caverzasi, E., Gallegati, M., Kinsella, S., & Stiglitz, J. E. (2016). Agent based-stock flow consistent macroeconomics: Towards a benchmark model. *Journal of Economic Dynamics and Control, 69*, 375–408.

Chang, M. H. (2011). Entry, exit, and the endogenous market structure in technologically turbulent industries. *Eastern Economic Journal, 37*(1), 51–84.

Chen, S. H. (2012). Varieties of agents in agent-based computational economics: A historical and an interdisciplinary perspective. *Journal of Economic Dynamics and Control, 36*(1), 1–25.

Colander, D., Goldberg, M., Haas, A., Juselius, K., Kirman, A., Lux, T., & Sloth, B. (2009). The financial crisis and the systemic failure of the economics profession. *Critical Review, 21*(2–3), 249–267.

Cristelli, M., Pietronero, L., & Zaccaria, A. (2011). Critical overview of agent-based models for economics. *arXiv preprint arXiv:1101.1847*.

Delli Gatti, D., Desiderio, S., Gaffeo, E., Cirillo, P., & Gallegati, M. (2011). *Macroeconomics from the Bottom-up* (Vol. 1). Springer Science & Business Media, Springer, Milano.

Diao, J., Zhu, K., & Gao, Y. (2011). Agent-based simulation of durables dynamic pricing. *Systems Engineering Procedia, 2*, 205–212.

Dosi, G. (2012). *Economic coordination and dynamics: Some elements of an alternative "evolutionary" paradigm* (No. 2012/08). LEM Working Paper Series.

Dosi, G., Fagiolo, G., & Roventini, A. (2010). Schumpeter meeting Keynes: A policy-friendly model of endogenous growth and business cycles. *Journal of Economic Dynamics and Control, 34*(9), 1748–1767.

Dosi, G., Napoletano, M., Roventini, A., & Treibich, T. (2017). Micro and macro policies in the Keynes+ Schumpeter evolutionary models. *Journal of Evolutionary Economics, 27*(1), 63–90.

Fagiolo, G., & Roventini, A. (2012). Macroeconomic policy in DSGE and agent-based models. *Revue de l'OFCE, 124*(5), 67–116.

Fagiolo, G., &Roventini, A. (2016). Macroeconomic policy in DSGE and agent-based models redux: New developments and challenges ahead. Available at SSRN 2763735.

Farmer, J. D., & Foley, D. (2009). The economy needs agent-based modelling. *Nature, 460*(7256), 685.

Goodfriend, M., & King, R. G. (1997). The new neoclassical synthesis and the role of monetary policy. *NBER Macroeconomics Annual, 12*, 231–283.

Guerci, E., Rastegar, M. A., & Cincotti, S. (2010). Agent-based modeling and simulation of competitive wholesale electricity markets. In *Handbook of power systems II* (pp. 241–286). Berlin/Heidelberg: Springer.

Hämäläinen, R. P., Mäntysaari, J., Ruusunen, J., & Pineau, P. O. (2000). Cooperative consumers in a deregulated electricity market—Dynamic consumption strategies and price coordination. *Energy, 25*(9), 857–875.

Holland, J. H., & Miller, J. H. (1991). Artificial adaptive agents in economic theory. *The American Economic Review, 81*(2), 365–370.

Kay, A. (2011). UK monetary policy change during the financial crisis: Paradigms, spillovers, and goal co-ordination. *Journal of Public Policy, 31*(2), 143–161.

Keynes, J. (1956). M. 1936. *The general theory of employment, interest and money*, pp. 154–6.

Kirman, A. (2010). The economic crisis is a crisis for economic theory. *CESifo Economic Studies, 56*(4), 498–535.

Kirman, A. (2016). Ants and nonoptimal self-organization: Lessons for macroeconomics. *Macroeconomic Dynamics, 20*(2), 601–621.

Kok K, Derzsi Z, Gordijn J, Hommelberg M, Warmer C, Kamphuis R, Akkermans H (2008) Agent-based electricity balancing with distributed energy resources, a multiperspective case study. In: HICSS '08: Proceedings of the 41st Annual Hawaii International Conference on System Sciences, IEEE Computer Society, Washington, DC, USA.

Krugman, P. (2009). How did economists get it so wrong? *New York Times, 2*(9), 2009.

LeBaron, B. (2006). Agent-based computational finance. In *Handbook of computational economics*. Elsevier, Amsterdam, North Holland (Vol. 2, pp. 1187–1233).
Müller, M., Sensfuß, F., & Wietschel, M. (2007). Simulation of current pricing-tendencies in the German electricity market for private consumption. *Energy Policy, 35*(8), 4283–4294.
Newman, M. E., Strogatz, S. H., & Watts, D. J. (2001). Random graphs with arbitrary degree distributions and their applications. *Physical Review E, 64*(2), 026118.
Rand, W., & Rust, R. T. (2011). Agent-based modeling in marketing: Guidelines for rigor. *International Journal of Research in Marketing, 28*(3), 181–193.
Rixen, M., & Weigand, J. (2014). Agent-based simulation of policy induced diffusion of smart meters. *Technological Forecasting and Social Change, 85*, 153–167.
Romer, P. (2016). The trouble with macroeconomics. *The American Economist, 20*, 1–20.
Roop, J. M., & Fathelrahman, E. (2003). Modeling electricity contract choice: An agent-based approach. In *Summer study for energy efficiency in industry*. New York. http://gridwise.pnl.gov/docs/pnnl38282.pdf, downloaded, 29, 2005.
Rumley, S., Kaegi, E., Rudnick, H., & Germond, A. (2008). Multi-agent approach to electrical distribution networks control. In *Computersoftware and applications*, COMPSAC '08. 32ndAnnual IEEE international, 28 2008-aug. 1 2008 (pp. 575–580).
Sanchez-Cartas, J. M. (2018). Agent-based models and industrial organization theory. A price-competition algorithm for agent-based models based on Game Theory. *Complex Adaptive Systems Modeling, 6*(1), 2.
Schelling, T. C. (1978). *Micromotives and macrobehavior*. W. W. Norton & Company, New York.
Shubik, M. (1975). *The uses and methods of gaming* (pp. 49–116). New York: Elsevier.
Simon, H. A. (1987). Behavioral economics. In J. Eatwell, M. Milgate, & P. Newman (Eds.), *The New Palgrave* (pp. 221–225). London: Macmillan.
Stiglitz, J. E. (2011). Rethinking macroeconomics: What failed, and how to repair it. *Journal of the European Economic Association, 9*(4), 591–645.
Stiglitz, J. E. (2015). Reconstructing macroeconomic theory to manage economic policy. In E. Laurent & J. Le Cacheux (Eds.), *Fruitful economics: Papers in honor of and by Jean-Paul Fitoussi* (pp. 20–49). Basingstoke: Palgrave Macmillan.
Tesfatsion, L. (2003). Agent-based computational economics: Modeling economies as complex adaptive systems. *Information Sciences, 149*(4), 262–268.
Tesfatsion, L. (2006). Agent-based computational economics: A constructive approach to economic theory. In *Handbook of computational economics*. Elsevier, Amsterdam, North Holland (Vol. 2, pp. 831–880).
van Leeuwen, E., & Lijesen, M. (2016). Agents playing Hotelling's game: An agent-based approach to a game theoretic model. *The Annals of Regional Science, 57*(2–3), 393–411.
Von Hayek, F. A. (1937). Economics and knowledge. *Economica, 4*(13), 33–54.
Yu, J., Zhou, J. Z., Yang, J., Wu, W., Fu, B., & Liao, R. T., (2004). *Agent-based retail electricity market: Modeling and analysis*. Proceedings of the third international conference on machine learning and cybernetics, Shanghai.
Zhang, T., & Brorsen, B. W. (2011). Oligopoly firms with quantity-price strategic decisions. *Journal of Economic Interaction and Coordination, 6*(2), 157.

Chapter 2
Towards a Big-Data-Based Economy

Andrea Maria Bonavita

Abstract On the threshold of 2020, we find ourselves in the middle of an extremely chaotic social and market scenario but at the same time with countless opportunities for emancipation relatively to everything we have so far considered as traditional.

The redemption of "standards" is an irreversible process that goes through behaviours increasingly distant from the experiential logic and increasingly guided by those who hold the knowledge of how our behaviours change.

Keywords Big data · Economy · Data-driven Darwinism · Ethical implications · Marketing

Introduction

On the threshold of 2020, we find ourselves in the middle of an extremely chaotic social and market scenario but at the same time with countless opportunities for emancipation relatively to everything we have so far considered as traditional.

The redemption of "standards" is an irreversible process that goes through behaviours increasingly distant from the experiential logic and increasingly guided by those who hold the knowledge of how our behaviours change.

This is the market of the reviews. First we search and then forward, share and recommend. And the more we do it, the more accurately our profile is traced.

This is the market of induced need. We are more and more buying things that we do not really need (or better, we also buy those), but we are even more being directed by those who are able to build invisible and persistent chains of attitudes based on our behaviours.

Nowadays, it is required to have a profile for any entity you interact with. Once, the profile was our identity, a few data. Essential, like the ID. For a few decades we

A. M. Bonavita (✉)
Nexteria S.r.l., Milan, Italy

© Springer Nature Switzerland AG 2019

F. Cecconi, M. Campennì (eds.), *Information and Communication Technologies (ICT) in Economic Modeling*, Computational Social Sciences,
https://doi.org/10.1007/978-3-030-22605-3_2

have gone further and we have been catalogued in clusters (in some cases we are still) as top-value or low-value customers for example. And at the end of the 1990s, if you were a top customer, Omnitel P.I. immediately answered you from the call centre and you also had a dedicated team of customer care agents.

On the threshold of 2020, the cluster is almost obsolete. Who owns so much data is undertaking the study of individual behaviour and commercial proposition aimed not only at our profile but at our profile in that particular moment and with that specific promotional message based on our mood and on how much budget we have available compared to how much we have spent in the last 6 months in that product category.

The study of behaviours and the deep understanding of the human being in his deep individuality have generated a completely different approach to the market.

Big and unstructured data have revealed unimaginable business opportunities if only the computational skills have exceeded the adequacy.

Machines perform human tasks with crazy speed managing a huge amount of information incomprehensible for our brain. The hype of artificial intelligence has been transformed into an evolution path where technologies are able to completely replace human beings (such as robotic process automation or process mining).

The worst is that we have also considered (and are still convinced) that entrusting to the machines exquisitely human tasks could generate a better lifestyle.

Some have foreseen (but not consumers) that machine needs a lot of data and needs to be constantly fed by that data to operate properly.

Where did all this data come from? How are they produced? Who owns them and how does get them? Today's data is the new precious resource (see the case of Cambridge Analytica which I'll talk about later) and we are the mines and miners ourselves with the difference that we deposit this treasure inside machines that execute algorithms and that grind and retract our behaviours to make us live better through the almost total control of our environment.

On the other hand, we have equipped ourselves with a new sensory appliance, made up of apps, mobile devices and accessories, environmental sensors, data and algorithms that are developed and embedded in daily and professional life. A bold attempt to live in a way that is unprecedented in our history.

In 2020, more than 34 billion Internet of Things devices will create new ways of perceiving the reality that surrounds us.

I recently had the opportunity to be selected as Alexa's beta-tester before being released on the market at the end of the past year.

Now Alexa knows everything about me and my family. Thanks to our conversations and requests, Alexa has learned to better understand what we are asking for and now answers quite well. She plays relaxing music after dinner and tells us jokes. She manages the lighting in the rooms and adjusts the thermostat setting.

Amazon tells me to buy items compatible with Alexa and offers them at a good price because, after all, I do not really need them. But what I really pay are not digital coins, not euros. I'm paying with data, personal data. A lot of personal data.

We must be aware that the amount of information we throw up in the cloud is a great responsibility. Not just for how much and how we change the market's laws but for how the market owns us. The market of profiles is not new (just look at the Cambridge Analytica matter and the conspiracy dynamics that have arisen).

But Facebook is (still) a fully functional platform.

A few dozen "likes" can give a strong prediction of which party a user will vote for, reveal their gender and whether their partner is likely to be a man or woman, provide powerful clues about whether their parents stayed together throughout their childhood and predict their vulnerability to substance abuse. It's quite easy to understand your needs and future needs. And it can do all this without any need for delving into personal messages, posts, status updates, photos or all the other information Facebook holds.

The same is for every entity able to fetch data from the mass.

Cost and Opportunity: Why We Buy?

If we try to take the intricate path of mental accounting, we must bear in mind that every economic decision is made through an evaluation of cost and opportunity.

The cost of going to the quarter-finals at Wimbledon (I'm a tennis and King Roger fan) is what takes shape in my mind compared to what I could do with those 2000 euros.

And I would only do this expense if it were the best possible way for me to use that money, but not by limiting the consideration to the cost.

Is it better to buy a new dress?

Is it better to go abroad with my wife and daughter?

Is it better to save money for a crisis time?

How do I know which of the endless ways of using 2000 euros will make me happier and more satisfied? The problem to be solved is too complex for anyone and it is crazy to imagine that the typical consumer will get involved in this type of reasoning. Especially me.

Few people do this kind of business accounting. In the case of the quarters at Wimbledon, many people would consider only a few alternatives. I could comfortably watch all the matches including replays of the best shots sitting comfortably on the couch and use that money to make my daughter attend about 20 ski lessons.

Would that be better?

To better understand how it works the mental process that leads to the purchase, or rather, to the decision to buy a certain good, we must distinguish between purchase utility and transactional utility.

The utility of purchase is that PLUS that remains after we have measured the utility of the object purchased and then subtracted the opportunity cost of what has been given up. From an economic-financial point of view, there is no value beyond the acquisition value.

If I am really thirsty, a two-euro bottle of water sold directly to the tennis club is the best thing I could have from the point of view of utility. Realizing that with those two euros I could have bought four at the supermarket, in a consistent process of mental accounting, should make me think about waiting because the objective evaluation of the price overrides the immediate need. If, for the same price (2 euros), I were offered a four seasons pizza, the case should be similar. Unfortunately I am not hungry but I am thirsty and very thirsty. Now.

To tell the truth, we also give weight to another aspect of the purchase: the perceived quality of the deal that is proposed to us, an aspect that is captured by the utility of the transaction. This is defined as the difference between the price actually paid for the item and the price you would normally expect to pay (i.e. the reference price).

Imagine you are on the central court looking at Roger and there you buy a bottle of water (the same bought at the club). It's very hot and we're in ecstasy in front of Roger but the price of that bottle is too high and produces a negative transaction utility: in other words, we think it's a "scam". On the other hand, if what you paid is below the reference price, then the transaction utility is positive: it is a "bargain", as if the ticket for the quarters at Wimbledon were offered at 1.500 euros.

In fact, it happens that we buy that bottle for seven pounds.

One thing is transactional pleasure and satisfaction. Another thing is the concept of usefulness of the good and possession.

Those who use the data wisely know how to trace some facets of our behaviour that direct more towards one type of pleasure than another.

The black Friday is the most obvious example of data-driven-manipulation economy.

Yesterday I had a look at a well-known brand sports smartwatch purchased last black Friday, which I used up for a couple of months. Now in a drawer. I wondered on the basis of what mental process I was induced to complete that purchase and I could easily understand that both transactional and asset use elements intersected.

In short, that smartwatch is now in the closet and (1) I'm not using it anymore but (2) I'm still convinced I bought it at a great price and made a bargain. I wonder why, however, I feel a strange sense of fluctuation between transactional complacency and actual satisfaction related to possession.

Almost as if the awareness of the poor arguments on the usefulness and preponderance of a positive shopping experience have generated a cognitive bias.

Since the transaction utility can be both positive (the bargain of life) and negative (a powerful scam), it can either prevent purchases that would increase our well-being or induce purchases that are just a waste of money.

Considering those who live in comfortable environment, the usefulness of negative transactions can prevent us from having particular experiences that would provide happy memories throughout our lives, when the amount of the overcharge paid would be long forgotten. The idea of achieving good deals can, on the other hand, encourage us to buy items of little value. There is no one who does not have a smartwatch like mine in their drawers but who considered it a real bargain to buy it at a particular time simply because the price was very low.

Just like the smoker who doesn't quit smoking, we are suffering of cognitive dissonance. We know that a good is unnecessary and we are inclined to justify a weak utility through a positive transactional experience. The problem is that we do not realize that we have appeared in the film of the economy where the screenplay is written by those who know how to guide our behaviour through the indiscriminate and massive use of data.

And since the overwhelming majority has the mindset black Friday branded, the seller has an amazing incentive to manipulate the perceived reference price and to create the illusion of "bargain".

The messages that induce people to buy are silently deafening and generate a state of exhaustion in which people do not have enough willpower to resist the temptations of discounts, losing the cognitive faculties necessary to elaborate complex decisions.

Data-Driven Evolution: Data-Driven Darwinism

The volumes of the coffee compatible capsules of a well-known brand are staggering; the demand is extremely high.

Officially established in 1998 from the merger of Rondine Italia, a pot producer, and Alfonso Bialetti & C., Bialetti has seen an unstoppable growth at international level over time, achieving a series of goals through investments and acquisitions, and then had its debut in the Stock Exchange in 2007, with a 74% share of the coffee maker market.

In 2015 the first economic difficulties began: the first debt with the banks was to create a series of points of sale, initially only in shopping centres and then also for the main streets of the city, in addition to the production of coffee capsules, a phenomenon that in those years was increasing powerfully in Italy. The project, however, is not successful.

Sales continued to fall, with a financial indebtedness of 78.2 million euros in 2017, compared to net equity of 8.8 million euros, and a loss of 5 million euros, compared to a profit of 2.7million euros in 2016. The debt agreement expires, the stock market price is revised downwards, and the Group has been facing a loss of around 80% since 2007.

Today we talk about the risk of bankruptcy and uncertain future, so much so as to lead the company to "the impossibility of expressing an opinion on the consolidated half-yearly financial statements at 30 June 2018". Elements of uncertainty were "already indicated in the report on the financial statements prepared by the Board of Directors, which may give rise to doubts about the company continuity".

5.3 million euros lost in the first half, a 12.1% decline in consolidated revenues, for the disappointing amount of 67.3 million euros in total revenues. This is the situation reported by the Group, an outcome mainly due to the "contraction in consumption recorded on the domestic and foreign markets", as well as to the situation of financial tension, "which caused delays in the procurement, production and delivering of products for sale both in the retail channel and in the traditional channel, leaving significant quantities of backorders in the latter channel".

Bialetti, do you know the brand of moka pot? Exactly them.

Bialetti has invented a product. You must keep in mind that before the moka, the coffee was made through a process of infusion: there was the Neapolitan moka that, once the water boiled, had to be turned over and, due to the effect of gravity, the water passed through the filter and, due to the effect of the infusion, the coffee was ready.

Then Bialetti comes and invents a device that transformed the market, a pleasure for all the senses.

Practical of course, but the success was made by the spectacular sound and the scent that radiated throughout the house. You couldn't have got the MokaBialetti.

Then come the pods with prices that are close to 70 € per kg compared to the 3 € of the moka, but the end user does not care how much it costs and the market moves to the pods.

Bialetti, in a booming economic position, should have and could have converted because this market has been changing for about 20 years (Nespresso was established in 1986 in Switzerland and now has 700 stores and 13.500 employees).

Lavazza catches the ball and follows the steps of Nespresso.

Illy is trying; the others are struggling; many will end up like Bialetti.

A story like many others.

No one says it's easy to convert a company, but the missed conversion was an announced death. Bialetti tried to produce espresso machines but using generic capsules (tea-type filters) or proprietary pods hoping to sell machines and coffee, but it was a huge flop as expected.

They have not aimed at the capsules of the two main brands and have paid for the consequences; today the pods market is firmly in the hands of Nespresso, followed by Lavazza.

Bialetti would have to make agreements with the two sellers and create a machine for their products. This was a way, but how could we blame them? Who can say what was right and what was wrong?

"Netflix? It has no chance", said the CEO of Blockbuster. Silent innovation, the creation of new behaviour and the introduction of an induced need have shifted the market.

"Neither RedBox nor Netflix are remotely on our radar in terms of competitiveness", declared Jim Keyes in 2008.

"Five hundred dollars? Paid by instalments? With a contract? I think it's the most expensive phone in the world", said the former CEO of Microsoft about the first iPhone.

"And people who use it for work won't like it because it doesn't have a keyboard. Which makes it uncomfortable to send emails".

"Mobile games are a passing phenomenon, from the customer point of view" said in 2011 by the President of Nintendo North America, Reggie Fils-Aime.

This is the screenshot of Facebook in 2004.

It was created as a website dedicated to Harvard students. A directory to find classmates. In 2005 Facebook opened to other institutions and began the compulsive behaviour of those who went to check every day if their school had been added (exactly the same behaviour that now insists on the single person sought). It works on the layout and the data that are shown.

In 2006 comes the first feed that is nothing more than a set of organized data.

In 2007, you need to be 13 years old to subscribe and the news feed was launched, so you can really make other business through data that users make available on the social network.

In 2009 Facebook lands on mobile and starts the real-time news feed. Other real-time data that look like bottles of water when it's so hot.

In 2011, the layout was completely reinvented and the most important innovation was the appearance of the timeline. The focus shifts from personal data to daily actions. Still data but organized over time.

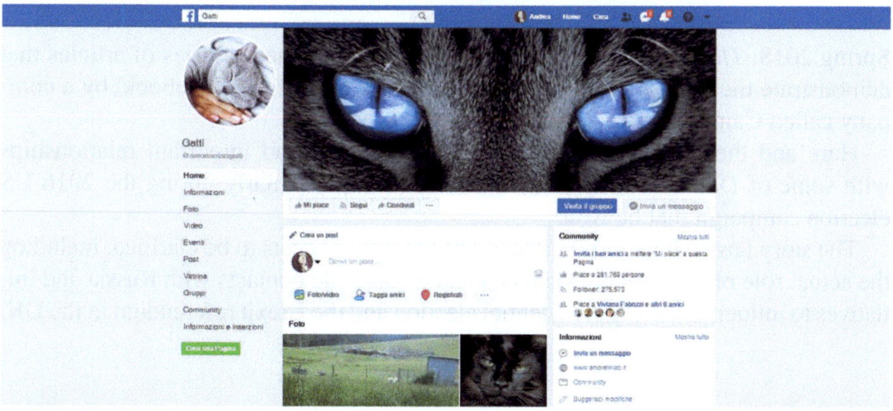

Nowadays, worldwide, there are over 2.32 billion monthly active users as of December 31, 2018. This is a 9 percent increase in Facebook year over year. This is compared to 2.27 billion for Q3 2018 (Source: Facebook 1/30/19).

There are 1.15 billion mobile daily active users for December 2016, an increase of 23 percent year over year (Source: Facebook as of 2/01/17).

Highest traffic occurs mid-week between 1 to 3 pm (Source: ZEPHORIA blog).

On Thursdays and Fridays, engagement is 18% higher (Source: ZEPHORIA blog).

Photo uploads total to 300 million per day (Source: Gizmodo).

Average time spent per Facebook visit is 20 minutes (Source: Infodocket).

On the other hand, Jeff Bezos was thinking about books, and in a garage in Seattle, he created something that was initially called Cadabra: because of a pronunciation not so auspicious, he gave a new name to what we now know as Amazon.

How can we identify differences and similarities between the history of Amazon, that of Facebook and that of Bialetti?

I think it's enough to quote that today when you say "To be Amazoned" you mean "to sit back and watch helplessly while someone steals customers and profits from your business".

Do you think "Amazon Prime Now" had a business plan or a behaviour plan that supported it? In the warehouse of Amazon Prime Now, there is a brass bell that rings when an order arrives that must be processed within an hour. Probably the same bell that rang at the time of the garage in Seattle when the first orders arrived on Cadabra.

Same questions, same doubts, same awareness for those who govern us tacitly today.

But what does Google, Apple, Facebook and Amazon have in common? Why are these four companies dominating the market and coining a new market? Because they have skillfully and unscrupulously used endless amounts of data. They have traced and studied behaviours, eliminated mathematical averages and opened the way for micro-marketing through transactional baits. And who will be the first of the giants to become extinct?

Then Cambridge Analytica comes...

The September 11th of the Data

Spring 2018. *The Guardian* and *New York Times* published a series of articles that demonstrate the misuse of a huge amount of data taken from Facebook, by a company called Cambridge Analytica.

Here and there, it says that Cambridge Analytica had important relationships with some of Donald Trump's closest associates, particularly during the 2016 US election campaign that he won.

The story has endless ramifications and there are aspects to be clarified, including the actual role of Cambridge Analytica and its possible contacts with Russia and initiatives to influence the US presidential election and the Brexit referendum in the UK.

On September 11th of the Twin Towers, we saw the black swan. We realized that enormous amounts of information on behaviour did not anticipate behaviour such as that of hijackers and their instigators.

The September 11th of the data didn't also allow us to realize that we are daily subject to a legitimate violation of our will.

Spring 2018. The title Facebook collapses. The controversy over the use of the data reopens due to an informer. We are used to hearing of informers in the criminal sphere, in mafia. But this time we are talking about an informer of something different. Christopher Wylie, a former employee of Cambridge Analytica, decides to tell his past to *The Guardian* and delivers a heavy blow to Mark Zuckerberg. Social networks use their users' data and we do most of the work. A sharing today and a like tomorrow: here the famous data points are, where our profile is rebuilt.

This is what the application called "thisisyourdigitallife" did, created by Aleksandr Kogan. To use this application you had to login through your Facebook profile.

It's important to keep in mind that at that time Facebook policy was still very permissive and external apps could use users to collect information about their contacts. The conditions written in small on which you always click "ok" to go ahead contained this information.

When the Facebook policy became more austere, the app developer decided to violate the new terms of use and sold the collected data to Cambridge Analytica. This type of sale is prohibited by Facebook which must suspend the account and sanction. Wylie claimed that Facebook was aware of this data traffic but did nothing. The Cambridge Analytica account was still active on Friday, March 16, 2018.

Personality traits are difficult to identify from a profile but once extracted they can be used for trivial things such as induce people to vote for a political party or to buy something through perfectly targeted messages.

In a presentation at the 2016 Concordia Annual Summit in New York, Mr. Alexander Nix discusses the power of big data in global elections. Cambridge Analytica's revolutionary approach to audience targeting, data modeling and psychographic profiling has made them a leader in behavioural microtargeting for election processes around the world.

Speaker: Mr. Alexander Nix CEO, Cambridge Analytica

Whoever chose Abercrombie & Fitch – explained Wylie – tended to be less moderate and more liberal. Wranglers, on the other hand, were usually very conservative about the concept of "order". And again: those who followed *Vogue* magazine or Macy's warehouses tended to be more liberal and outgoing. One of the things Cambridge Analytica noticed extracting Facebook data – Wylie said – was that fashion brands were extremely useful in producing algorithms that indicated users' thoughts and feelings.

Therefore, it has not only collected information, but it has also used it to develop its models, which were also used during the 2016 presidential elections.

This story, which ends with the bankruptcy of Cambridge Analytica and a serious blow to Facebook, was discovered thanks to the revelations of an informer. An informer of the "data".

Ethics, Not Metrics

So when we talk about big data and big-data economy, we refer to an extremely complex system of relationships where our daily life and habits write the script but also recite it.

It is a system we try to give rules, but these rules are not sufficiently dynamic at the same evolution level of the system itself. It is not enough to define a policy; it is not enough to have a guarantor to protect the data. We all need extreme awareness and perhaps a less unscrupulous approach that has a founding ethic baseline.

The increasingly massive and conscious use of the information extracted from the infinite amount of data, besides guiding market logics, can also allow humanity to extract useful meaning for global cultural growth.

In a world where in 1 hour more words are "tweeted" than those present in all the texts written in Ancient Greece, an ethical rationalization of knowledge becomes perhaps fundamental for the entire human race, allowing us to better know ourselves and to place ourselves in a market that is populated by billions of intersections every second.

Thales of Miletus is a Greek philosopher who lived around 600 BC. Aristotle, in his *Metaphysics*, dealing with the philosophy of the origins, defines Thales as the "initiator of this type of philosophy". Thales is considered the first philosopher in the history of Western thought.

Philosopher but also scholar and forerunner. Before him, the only way to explain natural events was to refer to legends about divinities. He was the first to introduce the idea of using data and intelligence to explain natural phenomena.

He collected data regardless of whether they could be used immediately. He just collected them. But above all, he had never been to Wimbledon and had never paid 7 pounds for a bottle of water. Perhaps he was moved exclusively by the sense of science and the value that his studies could bring to humanity. He was not influenced by the reference price and had never experienced a black Friday. But he was collecting data.

And like all precursors Thales was heavily criticized, discredited and boycotted by his contemporaries.

Many blamed him for never having exploited his intelligence and Thales replied that the issues he dealt with were interesting regardless, without necessarily being the object of profit.

"...Since, poor as he was, they reproached him for the uselessness of philosophy, having predicted, on the basis of astronomical calculations, an abundant harvest of olives, even in the middle of winter, even though he had little money at his disposal, he took over all the oil mills of Miletus and Chios for a derisory sum, since there was no demand for it; When the time came for the harvest, urgently searching for all the available oil mills, he rented them at the price he wanted to impose, thus raising many riches and showing that it is very easy for philosophers to get rich, but nevertheless they do not care about this" (Aristotle, *Politik*).

If he had been an ancestor of Bezos, he would probably have had the wit to buy other oil mills, and obol after obol, he would have made nilo.com.

After all, Thales simply found a way to monetize the data but his approach was sincere and not driven by greed.

Chapter 3
Real Worlds: Simulating Non-standard Rationality in Microeconomics

Giuliana Gerace

Abstract In this chapter, the differences between the standard notion of rationality, used in neoclassical economic model, and the notion of "non-standard" rationality are highlighted. The notion of non-standard rationality has been used in an attempt to address the discrepancies found between the ideal cognitive attitudes of the homo economicus and limited rational abilities of real decision-makers. By contrast, this notion is not supported by any adequate nor exhaustive theoretical account of "non-standard" preferences and therefore doesn't provide useful applications for real-world simulations in microeconomics.

Keywords Rationality · Conventional economics · Behavioural economics · Economic preferences · Market simulation

Introduction

The following considerations describe main theoretical and application differences between the standard notion of rationality, used in neoclassical economic model, and the notion of "non-standard" rationality, which emerged in the frame of more recent behavioural economic experiments. The notion of non-standard rationality has been used in an attempt to address the discrepancies found between the ideal cognitive attitudes of the *homo economicus* and limited rational abilities of real decision-makers. By contrast, this notion is not supported by any adequate nor exhaustive theoretical account of "non-standard" preferences and therefore doesn't provide useful applications for real-world simulations in microeconomics. In this frame, an alternative theoretical perspective is introduced as a result of previously presented material. The contribution is purely conceptual and is intended to provide a possible basis for alternative simulations of decision-making environments.

G. Gerace (✉)
Università di Pavia, Pavia, Italy

© Springer Nature Switzerland AG 2019
F. Cecconi, M. Campennì (eds.), *Information and Communication Technologies (ICT) in Economic Modeling*, Computational Social Sciences,
https://doi.org/10.1007/978-3-030-22605-3_3

The Notion of Rationality: From the Neoclassical Model to Behavioural Economics

The understanding of rational processes underpinning human motivations is a central objective of the discipline of economics, insofar as this is a social science, namely, basically dealing with people agency and choice. The basic assumptions in this regard have long been grounded in the theoretical perspective, according to which individuals are self-interested and rationally behave in order to satisfy their preferences. This is the well-known neoclassical rational choice theory, which moves from some of the classical premises stated by philosophers/economists from the second eighteenth century. In the classical approach, inaugurated by Adam Smith (1759) and then consolidated by famous theorists as David Ricardo and John S. Mill, the knowledge of psychological foundations of individual behaviour in social and economic contexts rested on the assumption that individuals' selfish desires played a large role in promoting public interest (Marshall 1890). The neoclassical perspective narrowed this dominating view of human motivations (in order to simplify the model of economic interaction) by assuming that consumers of goods and services, i.e. firms and households, were to be considered the only actors of the economic playground and that, in their choices, they were basically profit-maximizing subjects. Despite its usefulness and comforting character, the rationality axiom of the neoclassical approach has not been able to contrast the challenges imposed by experimental evidences, which emerged in the last decades, in regard to decision-making processes of individuals in interaction.

The View of Conventional Economics

Neoclassical economics has long dominated the sphere of microeconomics since the late nineteenth century, so that it became a conventional approach in the field (Samuelson 1948). This view rests on the crucial argument that the economics discipline must be value-free, namely, it has to make no reference to human ethical goals or values. In addition, it rests on specific assumptions about market actors, such as the fact that they behave (i.e. they choose) according to a "rationally coherent" system of preferences. Precisely, according to this rationality axiom, the so-called *homo economicus* chooses in view to maximize his utility, namely, his self-interest (well-being or "profit"). On this basis, another important argument of conventional economics regards the possibility for the whole discipline to deduce its own conclusions from the above-mentioned specific assumptions about market actors. The aim of this conventional view is clearly to make the discipline of economics purely deductive, i.e. able to provide predictable results and therefore more scientific than any other social science. Importantly, the maximization premise of the neoclassical or conventional paradigm is to be linked to methodological individualism, namely, to the idea that the individual is the ultimate source of variables to be analysed. In this frame, it's sufficient for microeconomic theory to understand that individual behaviour is rational and self-interested in every economic activity, such as production, distribution, consumption and resource maintenance.

It has been argued (Levin and Milgrom 2004) that individuals' decision-making perspective in real world can essentially be explained by making reference to their preferences and the relating restrictions, i.e. the external negative incentives for them. In fact, on the one side, individuals act rationally in view to satisfy at best their preferences; on the other side, they act in condition of "scarcity": namely, they are realistically unable to satisfy all of their preferences at once and so they must make an "optimum" choice amongst various alternatives. Since all human conduct undergoes such "scarcity principle", which obliges to choose amongst alternative, all fields concerning individuals' interaction (also strategic interaction) represents a potential sphere of application of rational choice theory, which has to be therefore considered not as a discipline but rather as a method (Mathis and Steffen 2015; Becker 1976). Due to its high level of generality and abstraction, the optimization approach has been used to analyse not only households' choices about consumptions and savings or rather business decisions about investments or hiring; it has been also applied to people choices about education, migration and various other domains of sociology, ranging from crime to marriage (Becker 1976).

Despite the realistic image of individuals having to make an optimum choice amongst different alternatives, the assumptions dominating rational choice theory are clearly ideal, namely, they refer to stability and uniformity factors, which are unlikely to hold in real worlds. First of all, individuals as rational utility maximizers are in principle able to make inferences on the basis of perfect information, as if they could carefully weigh costs and benefits of a specific choice; therefore they can hold a set of consistent beliefs (certainty, risk, uncertainty) about the outcomes resulting from their actions or from other actors' actions, which can be easily inferred and calculated. Moreover the "rational" preferences driving their maximizing behaviour are assumed not only to be consistent to some reference criterion but especially stable in time (i.e. the decision-maker has the same preferences about future plans at different points in time) and unambiguously ranked. Namely, the hypothetic individuals' desires in relation to outcomes, the so-called utility functions, can be easily deduced (calculated), starting from an identifiable reference point, since they are time consistent (not subject to changes over time), environment consistent (not subject to influences by any environmental factor) and nonconflicting (there's no conflict between two or more opposite individual motives, e.g. between short- and long-term goals). Importantly, both components of rational choice assumptions, beliefs and preferences, are perfectly compatible with an optimization paradigm independently of their content.

The ideal character of rational choice premises (which is also the reason why the theory can be easily abstracted, formalized and therefore so flexibly applicable) rests on a crucial conception of rationality, which is extremely uniform and stable: decision-making procedures are rational insofar as consistent; in addition and more importantly, such consistency is uniformly assumed for all rational individuals, i.e. rationality is assumed to be the uniform inner structure of any individual decisional attitude; people are narrowly self-interested and consistently subject to this optimization-based principle—this is the only rational way to be.

It's also important to consider how, according to the conventional rational choice theory, individual beliefs and relating inferences derive from the sole observation of the environment. Particularly, in strategic interactions, jointly optimal equilibria are possible thanks to individuals' common knowledge of rationality (actors know that all other actors are rational optimizers) deriving from mutually observable actions. In particular inferences about other actors, beliefs/preferences are determined by belief about each actor's structural position (Sun-Ki Chai 2001). While inference processes are uniformly based upon axioms of formal logic and probability calculus, all variations in beliefs are caused by variations in access to observable information about the environment. Observability is substantially a function of the environment and not of the individual's mental characteristics. This implies that mental states are not taken into consideration as independent variables of rational inference (and predictions), and neither do they need to be investigated as a possible determination of rationality.

In general the hypo-deductive character of conventional rational choice paradigm entails a positive aspect of the theory, which is predictability: empirical prediction can be made on the basis of individual utility functions. For years the calculability underpinning the ideal decision-making environment of rational choice theory led to the elaboration of perfectly efficient market model, which at least served as standard approximations to real-world scenarios (Sun-Ki Chai 2001).

However, in the last decades, empirical evidences demonstrated that there are sufficient limits in the way standard rational choice can elaborate a plausible model of a real decision-making environment: conventional assumptions not only lack verisimilitude in many circumstances (e.g. in regard to time consistency of preferences) but also fail to accurately predict a wide range of human behaviours. It has been ascertained that individuals' behaviour in real worlds feature perspectives going far beyond the narrow, simple axioms of standard rationality, so that both the theoretical parsimony and the generality of the conventional belief-desire model are no longer capable of explaining the complexity of human motivational and decisional attitudes.

The Rise of Behavioural Economics

In 1953 Maurice Allais exposed one of the first evidences of conflicts in the axioms of conventional economic theory (Allais 1953). Such evidence has been defined an anomaly (the Allais paradox), since it lacked the backing of plausible explanatory account. It showed inconsistencies in individuals' choices with reference to different situations (gambles): such inconsistencies clearly violated the independence principle of expected utility (if a person prefers choice A over B, then she should also prefer the same choice when these are mixed with some other event at the same probability).

Other evidences for deviations from the conventional view (Samson 2016) mainly suggest that individual decisional attitudes diverge from the optimization principle, showing often actions against one's own self-interests and rather triggered by a wide range of emotional factors and "bias". In particular, on the one side, preferences differ quite widely from the ideal of mono-maximization, i.e. the desire directed to a good or state of affairs useful for oneself, incorporating such factors as altruism and sense of justice. On the other side, the observability principle at the basis of beliefs formation in the conventional paradigm is contradicted by the fact that quite normal individuals are capable of holding beliefs that are not logically derived from or even contrary to observable evidence. In this regard, in the second half of twentieth century, a new field of study emerged, challenging the conventional model of stable preferences and beliefs as the only relevant attitudinal variables for determining actions and trying to provide theoretical account for more sophisticated model of human motivations in decision-making. Precisely, behavioural economics has risen as a subfield of microeconomics with the aim of studying human conduct in market and/or strategic interactions through field or laboratory experiments (Camerer et al. 2004). Rather than making purely logical assumptions, it relies on scientific experiments relating to concrete individuals' behaviour in view to justify deviations in equilibrium theory and to build new decision-making models in economic contexts (Weber and Dawes 2010). Deviations from the conventional notion rationality are introduced as "non-standard". Arguments in support to such new models of non-standard rationality derive also from neighbouring disciplines, such as psychology, social sciences, cognitive sciences and neurosciences. On the whole, behavioural microeconomics focuses on departures from the conventional assumptions of selfishness, time consistency, rational inference and strategic sophistication with applications to markets, policy and strategic interaction.

The first main concept reflecting the importance of a psychologically informed economics was that of "bounded rationality" (Simon 1982). This notion focused on the need for individual decision-making processes to be understood in relation to the environment but also to the limits of human cognitive and computational abilities: decisions are not always optimal but "satisfying", exactly because restrictions are not only in the environment but also in individuals' information processing. This led to the main thesis that individuals acting in the market with incomplete knowledge and information can still satisfactorily reduce their uncertainty and take decisions (Simon 1982; Kahneman 2003).

Important studies built on the notion of bounded rationality (Gigerenzer and Goldstein 1996) proposed that the rationality of a decision principally depends on structures found in the environment. People are ecologically rational when they make the best possible use of limited information-processing abilities, by applying simple and intelligent algorithms that can lead to near-optimal inferences: heuristics.

Nonetheless the main support to the thesis of limited human rationality derived from studies on heuristics and biases by D. Kahneman and A. Tversky since the early 1970s (Tversky and Kahneman 1974; Kahneman and Tversky 1979), which also provided important methodological contribution to the field of behavioural economics, because they advocated an experimental approach in view to understand economic choices made under different conditions. Along with emblematic experimental findings, their research led to the definition of a number of cognitive biases, which contributed to establish the idea of systematic and predictable "deviations" from standard rationality.

It has been considered (Della Vigna 2009) that the most important systematic bias is the so-called framing effect, encompassing other fundamental cognitive biases. Framing effect refers to the way people tend to favour specific alternatives when they are represented (framed) positively and reject them when they are framed negatively: for example, with reference to information gaining about the employment rate, the datum of 95% employment tends to be preferred to the datum of 5% unemployment. Such cognitive disposition also implies what has been defined the "loss aversion" (Kahneman and Tversky 1979), namely, the fact that individuals' decisions regarding potential earnings mainly feature risk aversion, while decisions regarding potential losses tend to feature risk seeking. This concretely challenges the basic tenets of conventional expected utility: basically, greater value is associated with losses than to equivalent gains; consequently, potential gains are often not realized in reality. Associated to loss aversion are the "status quo preference", regarding the fact that decisions maintaining a *status quo* are privileged over decisions of change associated with potential uncertainty, risk or potential losses; the "endowment effect" (Kahneman et al. 1990) regards the following disposition: although rationally speaking a person should ascribe the same value to a good irrespective of whether she possesses it or not, the monetary value she is disposed to accept in view to sell it is always higher than the monetary value, which she is disposed to pay in view to buy it (Della Vigna 2009). According to Kahneman and Tversky's studies, frequent heuristics are also associated with two main biases: the availability bias and the anchoring effect. The first one relates to judgements about the probability of an event, which depends on the perceived likelihood of the same event in the judging of a person's mind, namely, it concerns the overestimation of probabilities if an event is cognitively available. The anchoring effect also relates to probability judgement, which are normally made intuitively or with reference to an arbitrary anchored point of view: for example, I can mistake judgements by making inaccurate numerical estimations or basing on biased memory, because I am contingently making reference to a perceived image or notion.

In line with this field of study, a variety of experiments have been made, whose data revealed a "non-standard" way of processing information, able to challenge the empirical validity of conventional rational choice and expected utility theories (Rabin 2000; Rabin and Thaler 2001). Building on previous studies regarding systematic biases, M. Rabin developed a specific structure to indicate three main forms of deviations from conventional rationality (Rabin 2002a, b; Della Vigna 2009): non-standard decision-making, non-standard beliefs, and non-standard preferences. Non-standard decision-making classifies the particular way decisions are taken, in contrast to the smooth deductive calculus of the maximization paradigm. This category includes the already mentioned framing and general rules of thumb, which individuals' mind uses to reach decisions more quickly, e.g. the mentioned availability bias. It also includes the herding and identity effect, according to which individuals tend to act with gregarious behaviour, mimicking others' and conforming with the crowd (Akerlof and Kranton 2000), also in absence of social pressure: this happens even amongst professionals and financial market analysts (Rizzi 2009).

Non-standard beliefs instead classify the specific way probability judgement are formulated in relation to cognitive biases, such as the confirmation bias, by which individuals' mind tend to give greater weight to information confirming an already given hypothesis, while other information or data, on the contrary, are disregarded (Nickerson 1998); the overconfidence bias, by which judgement concerning the probability of an event are based on an overestimation about one's own capabilities (Akerlof 2002); and the law of small numbers, by which individuals use to formulate the incorrect belief that small examples can be statistically extended to the whole population (Tversky and Kahneman 1974; Rabin 2002b; Della Vigna 2009). Finally, non-standard preferences identify preference typologies, which explicitly contrast the maximization utility principle. Amongst these are time preferences: it has been shown that individuals tend to focus on the present and to undervalue the future; this deviates from neoclassical predictions and from the standard assumption on time discounting, namely, on individual higher evaluations placed on investments for future goods (Shane et al. 2002). Non-standard preferences are also identified with the so-called social preferences or externalities in preferences, which concern reciprocity behaviour and clearly contrast with the conventional assumption of rational selfishness. In fact behavioural experiments have shown (Rabin 1993; Fehr and Gaetcher 2000) that people are not always selfish, but often altruistic (in many cases caring about fairness) or even spiteful. Also game-theoretical approaches to social science demonstrated such an attention of people for social norms and fairness, even in terms of normative expectations from others (Bicchieri 2006).

Building on prospect theory (Kahneman and Tversky 1979), other studies relating to the notion of "mental accounting"(Thaler 1985, 1999) defined the view according to which people think about a value in view to act or decide in relative rather than absolute terms, e.g. they can derive pleasure not just from an object value but from the quality of the deal itself (e.g. the nature of a transaction). The core idea in this case is that notwithstanding money is always the same and has no labels; it may not only carry formal value in itself but also carry other values concerning money's origin or the intended use of it: people can frame assets in relation to other factors such as wealth and card payment (e.g. treated differently than cash).

In general, the behavioural approach has demonstrated that individual preferences in strategic or risk decision-making are unstable (Ariely et al. 2003). Decisions can be "distorted" by influences not related to the rational goals of the actor. Importantly, such influences can derive from our own emotions. Thanks to behavioural economic findings, the role of emotions in value perception and in defining economic decision-making began to be taken more into account with emphasis on the notion of emotional intelligence (Della Vigna 2009). Particularly, it has been demonstrated that reasoning can be more effective when based on emotions (Ariely et al. 2003).

All behavioural economic findings have suggested that human decisions and motivational attitudes are strongly influenced by contexts and external cues, both at cognitive and emotional level: thought is not "autonomous". This invites a reflection on the role of influence on individuals' cognitive processes in economic choices: other individuals and organizations (e.g. politicians, advertisers) can affect our decisions by setting a "frame", which provides emphasis of some details at the expense of others (Goodwin et al. 2014).

Importantly, what behavioural studies established is the idea of systematic and predictable deviations from standard rationality. Individual behaviours vary across time and space, being subject to cognitive biases, uncontrolled reactions to certain informational stimulus, social influences and emotions; this means that decisions are the result of cognitive processes, which are less deliberative, linear and controlled than the conventional view wanted us to believe (Frank 2006).

Usefulness and Applications of Behavioural Economics

Comparing with the narrow view of human nature and the lack of contextual awareness of conventional economic view (which failed to predict the financial bubble responsible for the global great recession, from 2007 onwards), behavioural economics features a wider cognitive perspective. It has been claimed that behavioural economics can increase the explanatory power of economics by providing it with more realistic psychological foundations "which go beyond established notions of rationality" (Akerlof and Shiller 2009). In this frame, different from conventional economic models, conclusions about real-world mechanisms are not reached by chains of deductive logic based on established assumptions (complex and formal mathematical models of action) but emerge from bottom-up approach of interaction simulation.

While conventional rational choice models focused on predictable environment and didn't account for the rich variety of human motivations and cognitions found in real world, models based on non-standard rationality are strongly based on individuals' attitudes and heterogeneity in decisional processes.

A Wider Perspective in Explaining Economic Decisional Processes

As mentioned, evidences show that individuals are often irrational in choice, i.e. they do not follow the patterns predicted by conventional models. Influence by conflicting preferences or even by others' preferences (preferences of preferences) in decision-making processes contributed to our understanding of cases in which specific cognitive and emotional responses are likely to mediate the relationship between preferences and choice, therefore generating the so-called default effects (Gigerenzer 2008), which are options immediately, not actively chosen. Psychology has provided theoretical justifications for such cases of individuals' judgements by accounting for the notion of dual system of thinking (Kahneman 2011). This notion refers to a double cognitive structure in individuals' mind, in which two thinking systems are reciprocally influencing. Precisely, System I consists of thinking processes that are intuitive, automatic and experience-based, while System II is more reflective, controlled and analytical. Particularly, System I influences judgements by means of mental contents that are easily "accessible" at cognitive level, and this is also the reason why it can be considered "home" for heuristics and belief bias (Kahneman 2011), the latter being the paradigm case for dual processes in reasoning. In their being belief-based, heuristic processes are fast and automatic, whereas analytical reasoning is sequential and deductive. Even heuristics that are apparently conscious in application have an automatic component (Kahneman an Frederick 2005).

On the whole, the dual process notion contributed to a new paradigm in cognitive studies, including a number of theories concerning the double structure of cognition (Evans 2008), which treat System I as a "cognitive unconscious"(Lieberman 2003), also incorporating emotional processes and motivations (Hassin et al. 2005; Velmans 2000).

Almost all theories conceive System I as typically associated with multiple implicit cognitive processes and has been claimed to be evolutionary older than System II (Evans 2008). It's especially associated with "intuition" and linked with experience, since intuitive processes seem to be more accurate in areas where we have gathered a certain quantity of data and which we are therefore able to respond to with reliable and fast feedback. Importantly, intuition has been characterized as highly context-sensitive, generated and activated by environmental cues: in line with Kahneman's theory, contexts and decision-making environment have attributed a direct role in the cognitive processes driven by both systems of thinking (Djiksterhuis and Nordgren 2006). By contrast, System II is associated with language as a form of thinking under intentional control, which appears to be more coherent and consistent than System I and therefore is deemed to have inhibitory role. This level of reasoning can engage in hypothetical thought via supposition and mental simulation. A common definition is that System II requires a working memory of limited capacity, different from System I. Nonetheless, such short-term memory capacity doesn't prevent its executive and inhibitory functions (Evans 2008). Otherwise, it has been emphasized that unconscious processes can control our behaviour without us being aware of it, also providing a superior effectiveness in decision-making (Hassin et al. 2005).

In this regard, it has been shed light on some valuable insights on the limits of reasoned decision-making (Ariely 2008; Dijksterhuis et al. 2006). Not only heuristics help in saving us information search costs, thanks to automatic processing and implicit discrimination (without them we would be frozen into indecision), but also emotions are likely to matter for decisions, especially economic decisions, since in most cases they are likely to back such automatic cognitive processes.

While the conventional view treats emotional factors as affecting logical reasoning and therefore as interference with good decision-making (intelligence is identified with the ability to rationally control emotions), the non-standard perspective, as mentioned, relies on the role of emotions in decision-making through the notion of "emotional intelligence" (Ariely et al. 2003). In fact it's not always true that decisions based on logical reasoning are better than those based on intuitions and emotions. While reasoning seems to be most effective for what concerns simple economic decisions, "affective" intuition can provide the best results in more complex economic evaluations carrying an overload of information (Ariely 2008). Experiments concerning purchasing (Dijksterhuis et al. 2006) demonstrated that, contrary to conventional assumptions, it's not always advantageous to engage in conscious or excessively reasoned deliberation before choosing. Emotions are much better than overthinking in assessing actual individuals' "real" preferences.

In general, it has been considered (Della Vigna 2009) how a variety of decision-making deviations from standard rationality (e.g. self-control problems, social preferences, projections biases) are at least partially mediated by emotional states.

As well known in social science, preference for social norms, even when triggered by social pressure, is also a channelled by emotions (Elster 1994). Social intuitionism (Haidt 2001) and general social norms compliance is often backed or triggered by emotional/affective states, at both personal and sub-personal levels (Colombo 2014).

In general "mood" has been proved to be the channel for variations in non-standard choices. Experiments showed (Lowenstein and Lerner 2003) that even minor mood manipulations (environmental influences; weather conditions) have a substantial impact on behaviour and economic decisions, such as stock market returns (Hirshleifer and Shumway 2003). If mood is the channel for these effects in choices, the question should not be what are the mood-altering events but what are "exactly" the cognitive mechanisms responsible for these variations.

Nonetheless, the way environmental decision-making affects our economic decisions through our unconscious thinking (System I), particularly our consuming choices, has been object of important behavioural studies. As mentioned, thought is not "autonomous": other individuals and organizations (e.g. politicians, advertisers) can affect our decisions by setting a "frame", which provides emphasis to some details at the expense of others. Basing on this idea, some part of the so-called new behavioural economics, organized under the concepts first of libertarian paternalism (Sustein and Thaler 2003) and later of Nudging (Thaler and Sustein 2008), intended to provide theoretical support to governmental policies. Instead of simply investigating the non-standard aspects of individual rationality, latest studies engaged in accounting the way in which market agents "can be induced" to behave more in accordance with norms of governmental economic policy (Thaler and Sustein 2008). The assumption behind this purpose is that there is a "correct" behaviour each market agent could actually adopt; nonetheless, their decisions are more than often distorted by cognitive bias, which drive their decisions far from practically correct behaviour. In this regard, increasingly strong focus has been put on the so-called decision-architecture (Camerer et al. 2003), namely, the possibility to design the environmental decision-making in a way to facilitate consumers with the best possible options and to avoid the effect that aversion of regret can have is consumer inertia.

We have briefly examined how the dual process perspective unveils an inner complex structure of human cognition, which is divided into two functionally different mental systems, that are nonetheless nested in each other: on the one side, System II controls and monitors the suggestions (thoughts and actions) of System I within the constraints of working memory and volitional capacity; on the other side, the logical inferences of System II must first base on the cognitive and/or motivational patterns that have been associated in System I. Both levels of thinking are someway responsible for our choices, in relation with our decision-making environment. On this basis, the non-standard paradigm has integrated the classical utility function with a more accurate analysis of "realistic" preferences in behavioural economic models.

Contributions to Real-World Simulations

Knowledge of the psychological foundations of economic agents is basic for real markets simulation. The bounded rationality of market individuals and firms has to be reflected in microeconomic models, which are able to account for agents' heuristics and also to reproduce settings within which since it's not possible to eliminate biases; hence, the effect of non-standard behaviour aggregates in a complex manner or linearly (DellaVigna 2009). Similar models have to be able to explain behaviours such as those leading to stock market swings or the interaction between consumers with biases and profit-maximization firms. Particularly, an adequate modelling of consumer behaviour is basic to the discipline of marketing: non-standard features can have a disproportionate impact on market outcomes.

Conventional microeconomic models, such as econometric and dynamic stochastic equilibrium models, have been questioned since the advent of the global financial crisis. In fact they were able to predict results concerning business as usual states of affairs, but they failed to predict the crash by assuming a perfect world and by accounting for minimal fluctuations around an equilibrium state, which doesn't exist in case of an economic crash. Therefore, the global financial crisis has reminded us the economic role of social capital components, such as trust and confidence, which are difficult to incorporate in disembodied models of rational decision-making.

It's growing the interest towards experimental simulations of non-standard behavioural attitudes in economic setting, such as herding amongst investors (Secchi and Neumann 2016) or the degree of trust in market by economic agents, particularly by relying on agent-based modelling. This technology is able to simulate complex systems of interacting agents. Specific studies in agent-based economic models, i.e. agent-based computational economics (Tesfatsion 2002), took the advantage to explore how various micro-level assumptions interact to produce aggregate macro-level patterns (Page 2012).

In agent-based modelling, the simulation of real markets is based on the self-organized interaction between autonomous and adaptive agents. Namely, individuals (agents) are bounded rational entities, able to learn from experience and to change behaviour accordingly. They especially act autonomously and with heterogeneous perspectives with reference to the dynamic interacting environment. The most important characteristic of such adaptive multi-agent systems is their bottom-up approach: system behaviour emerges from the interactions of individuals, so the model focuses first on the design of agents' behaviour and interacting conditions. But, instead of assuming a fixed set of alternatives, the model is designed to generate alternatives, particularly alternatives for dealing with uncertainty that do not assume knowledge of probabilities (Windrum et al. 2007). So the state of markets is the result of emergent properties featuring, e.g. turbulence in the patterns of financial/industrial dynamics. The state of the economy is not necessarily an equilibrium, but rather a complex system evolving through time.

Agent-based bottom-up approach is extremely useful to understand the emergence of economical phenomena like price fluctuation. In the bottom-up approach, it is also relevant to consider the emergence of interaction networks: in real worlds agents form small-scale networks to exchange information and commodities; in turn, the network structures influence aggregate results, e.g. the market design.

As mentioned, the economic activity of bounded rational agents is essentially a social relationship, basically influenced by context-dependent factors: prices can be determined by trade settings and not only by the quantity of supply and demand (Simon and Voyer 2014). Therefore, agents rely on the one side on a degree of rationality, on the other side on a degree of opportunism, the latter being strongly influenced by emotional factors. In this frame, agent-based models of consumer behaviour open new possibilities in terms of the degree of realism and complexity of the simulation of markets.

At first the new assumptions were mostly incorporated in belief-desire-intention models (BDI), inspired by a certain perspective on practical rationality (Bratman 1987), whose architecture allows for the representation of deliberative and heterogeneously motivational attitudes of individuals. Precisely, BDI models link the rational consistency of agents' choices to their heterogeneous motivational attitudes and, contextually, to their freely chosen "instrumental" objectives. BDI models don't include any prescription on utility functions, namely, on how to sense and interact, e.g. in the marketplace (Baptista et al. 2014).

Also, a very crucial development for real market simulation is the modelling of agents' emotional and social processes. Importantly, social processes include the impact of norm compliance and can lead to individuals "emotional contagion" also in this regard, namely, in terms of conformity. Such "human-like" decisional attitudes in economics have only recently been explored in the so-called affective computing studies, which focused on the recognition, interpretation and simulation of human emotions in interactive contexts. These studies have recently led to the development of computational models (Rumbell et al. 2012) and specifically agent-based economic models (Espinosa. et al. 2017), which represent a further contribution to the development of simulation infrastructures able to reproduce human-like decision-making dynamics.

The Limits of Behavioural Economics

It has been argued that the statements supporting behavioural economic models are much more inclusive and close to reality than in the conventional view, but they are also much looser since cannot be treated in the same deterministic manner (Secchi and Neumann 2016). They provide less opportunity for developing mathematical models based on simple axioms about behaviour nor is behavioural science all that economists need to know: explanation or predictions of economic phenomena sometimes require inputs from other social sciences about individuals' cognitive and motivational attitudes, which are not always incorporated in both models and theories. On the whole, behavioural economic models don't express the ideal of "scientific" social science pursued by neoclassical economists.

Limited Predictability of Behavioural Economic Models

Multi-agent-based models allow for the simulation of emergent properties of economic markets, basing on agents' interactions that endogenously generate structural adaptation. Despite the versatile applications of such "virtual reality" simulations, some doubts its efficiency in predicting actual direction of change within an evolving market (Sharpanskykh 2007). The main concerns about the poor predictability of behavioural simulations derive from a certain approximation or alleged incompleteness of models' formalization. Many developed models represent only a limited number of (apparently) unrelated organizational aspects of the market. Furthermore, some of them make little use of a rich theoretical basis developed in social science. This may undermine the practical feasibility of such models (Sharpanskykh 2007).

Bounded rationality has been widely incorporated in economics but in models that mainly represent individuals' cognitive (or information processing) limitations as fixed and exogenous. Similarly the variety of preferences is often represented as stable lists, which means that they are exogenously defined, while more realistic models should allow for the representation of heterogenous but also endogenously generating preferences (Fisman et al. 2007). Actually the endogenous generation of both cognitive and motivational attitudes would be an efficient basis for the simulation of realistic emergent properties.

With reference to complexity in economic simulation models, bounded rationality poses some challenges in practice, such as the fact that its nature and extent vary across both decision-makers and problems (Diasakos 2013). Therefore, in order to derive realistic choices, the point is to exactly model not only individuals' interactions but also the interaction between agents and the precise structure of problems themselves. So our understanding of bounded rationality remains incomplete until we are able to explain, e.g. why an agent can depart from standard rationality in some problems but not in others and why different agents exhibit different cognitive limitations in a same given problem.

If behavioural economic models are only partially or not adequately able to explain and replicate existing reality, therefore it's unlikely that they will possibly and consistently predict future trends. In turn, if models formalization is incomplete, this must be due to a lack of accurate and realistic assumptions in theories, as we'll see in detail.

Limited Explanatory Potential of Behavioural Economics Assumptions

The notion of rationality in behavioural economic theories is surrounded by a certain ambiguity regarding its status as a descriptive or prescriptive hypothesis: it's not always obvious whether rationality is conceptualized in order to explain and predict how rational people act or in order to evaluate and prescribe the way people ought to act to be rational. In this regard, it has been distinguished between normative theories, i.e. theories of what is considered to be a rational choice (possibly to be conformed with), and descriptive theories, i.e. theories that just predict what people are actually able to do in various circumstances. In this frame, it has been argued that the prescriptive neoclassical notion of rationality is still essential as benchmarks on which to build descriptive theories, since it "characterizes optimal choices" (Thaler 2016). Nonetheless, we need a "new breed of descriptive theories" specifically designed for an understanding of how individuals actually behave (Thaler 2016).

Early behavioural theorists argue that behavioural economics is purely descriptive, in contrast to the normative authority of the neoclassical paradigm. However, in these theories, all behavioural departures from the predictions of utility theory have been interpreted as "anomalies" or errors: this clearly represents an extension of the normative notion of standard rationality. Indeed neoclassical expected utility theory has just been extended to incorporate findings from behavioural economics (Heidl 2016), so that rationality is still the key notion for the explanation of behavioural attitudes: both the reductionist focus and the benchmark of the neoclassical paradigm returns in disguise. The recent behavioural welfare economic theories (Thaler and Sustein 2008) also assume models of bounded rationality, which include the possibility of latent "correct" preferences *(*Heukelom and és Sent 2017), namely, the possibility for choice designers to estimate the optimal setting of options and for individuals to recognize the so-called optimal choices thanks to an inner preference structure. Indeed such theories suffer the lack of a consistent, objective and convincing theory of preferences (Hédoin 2015). In general behavioural economics is not supported by an adequate theory of preference, concerning both their generation and consistency. It doesn't include any explanation of the reasoning processes by which individuals construct their preferences, it doesn't justify any specific pattern of preference as "rational", nor does it provide any objective ranking of preferences at substantial level. Equally, in regard to social preference, neither game theoretical approaches in social sciences help to identify an adequate cognitive justification for individual social choices in interaction: the normative force of social patterns is often anchored to the utility function of Nash equilibria themselves (Bicchieri 2006; Binmore 2010). Similar perspectives have the merit to focus on individual motivational attitudes to the social space as crucial cognitive factor, also triggered by context-dependent variables. Yet, even assuming a plausible context dependency of individual preferences for social conformity, the specifics of how social contexts affect changes in individual preferences remain fairly vague (Paternotte and Grose 2012). Neither these approaches are able to justify normative attitudes occurring at unreflective level, namely, independently from any voluntary cognitive representation of norms.

In general, when individuals' decisions are inconsistent with the standard assumptions of rational choice, those decisions are treated as mistakes (Hausman 2012). So many context-dependent preferences revealed in experiments seem to be irrational; indeed none of these perspectives does answer the question of what is exactly to be considered rational. By contrast, our behaviour may be influenced by reasons (therefore preference consistencies) we are either not aware of, or we don't accept. Contrary to general assumption, behavioural anomalies may therefore matter (Infante et al. 2016). Interestingly, a number of recent critics to the tenets of dual process theories shed light on some theoretical fallacies, which contribute to the scarce explanatory potential of behavioural economic assumptions. Perhaps the most persistent fallacy in dual system theories is the idea that the automatic System I is responsible for all biased thinking, while the analytic features of System II cannot lead but to correct thinking (Evans and Stanovich 2013).

In general, in most of dual process theories of choice, the relation between intuitive and deliberative processes remains vague, so as the relation between internal processing in System I (e.g. between intuitive/associative processing and affection). Particularly, the description of the nature of System I requires further explanation and refinement (Martin and Sloman 2013). It's generally assumed that if System I is evolutionarily older than System II, then conscious reasoning only occurs under certain circumstances and always subsequent to System I processing (Kahneman and Frederick 2005). Otherwise (De Neys and Glumicic 2008), it has been shown the deliberative processes can be activated prior to intuitive processes and that sometimes that both systems of thinking process information in parallel, supporting simultaneous contradictory beliefs and preferences. It seems therefore that both systems necessarily cooperate in a complex interaction to produce choices (Martin and Sloman 2013). However, one interesting observation concerning the anteriority of System I regards the fact that, if its unconscious structure are activated first, then we should expect that this is home for the generation (let's assume the progressive formation) of the most reliable latent preferences, while the analytical rationality of System II would provide a decision support service (Infante et al. 2016). In this perspective, therefore, incoherent and non-selfish preferences should not be considered as deviations from some true (rationally consistent) underlying preferences, but the evolving preferences of a socially embedded agent. There would however persist the problem to explain how such preferences form and evolve.

As mentioned, differently than for System II, the nature of System I is quite uncertain.

Particularly, resolving the role of intuition, affection and perception with reference to automatic thinking is a critical issue for an adequate characterization of choices.

System I has been generally considered intuitive and associative, particularly operating at a fairly superficial level of "perceptual inputs": its cognitive functions are activated by the most salient options at stake, i.e. the ones that "stand out" relative to others (Dahr and Gorlin 2013). Particularly in this regard, since System I has been described as "perceptual" only with reference to sensory objects or event (they are capable of generating intuitive responses by means of their surface properties) (Dahr and Gorlin 2013), it's still to be explained why there can be evidence of attraction effects also with reference to abstract or numerical attributes of objects (Martin and Sloman 2013). Otherwise, System I has been viewed as basically affective and valuation-based (Slovic et al. 2007), namely, automatic choices or immediate preferences are triggered by an affective evaluation of the options. Nonetheless, again, the mechanisms explaining how such appropriate "intuitive" choices rise due to affective responses has been not accounted yet.

The general assumption concerning the cognitive structure of dual process theories is that one system is intentional, controllable and conscious, while the other one is unintentional, uncontrollable and unconscious. Yet there have been persistent objections to this rigidly distinctive typology. The crucial issue is that there are evidences for a misalignment of attributes between the two types of processing.

First of all, it has been shown that unconsciousness doesn't necessarily identify with inefficiency: research demonstrated that people are able to solve effortful mathematical problems unconsciously (Melnikoff and Bargh 2018). So although it's assumed that logical responding must originate from slow, deliberative System II processing and although it can be actually possible that belief-based heuristic responses have higher activation levels than intuitions, nonetheless there is evidence for the existence of intuitively generated logical responses (De Neys 2014) and for the fact that intuitive processing from System I can lead to right answers, while analytical processing from System II leads to biases in some circumstances (Stanovich et al. 2011): therefore the automatic processing of System I can cue a correct logical response from a reasoning process.

Another important misalignment has been detected with reference to the attributes unconscious/unintentional (Melnikoff and Bargh 2018). Indeed we can easily consider a number of cognitive phenomena that are intentional yet able to operate without conscious monitoring, especially in cases of routinized information processing (Logan and Crump 2010); many human skills in fact can be practiced unconsciously while remaining completely intentional activities: driving, typing and playing the piano (Bargh and Morsella 2008). In this regard, it has been observed that there are at least "two flavours" of intentionality (Melnikoff and Bargh 2018): on the one side, it's possible to consider intentional information processing, which are activated on purpose; on the other side, we consider intentional information processes, which are goal-dependent, i.e. they are activated by external stimuli but only in the presence of some relevant processing goal (e.g. breaking while driving or, at sheer motivational level, regretting for a previous choice, while performing a related decision).Such goal (or purpose)-dependent intentions are an in-between form of intentionality, which doesn't regard "purely unintentional" information processing (evaluating, stereotyping), neither concerns "purely intentional" information processing of the mind (e.g. analysing).

Also with reference to the relation between intuitive and deliberative information processing, a number of question should be addressed, particularly concerning the activation of deliberate analytical thinking and the different representation of the information involved in both processes. A crucial question is what is the source of an analytical response (contingent or hypothetical)? (Evans 2007). The task is to determine what triggers individuals' engagement in analytical processing, also in absence of some obvious situational or instructional cues. An argument considered is that both the "willingness" and the ability to engage in analytical reasoning are important components of human rationality (Pennycook 2017). There has to be something that spontaneously triggers analytic thought, basing on some (unconscious) volitional attitudes, while the cues that induce people to think analytically surely can differ. Again, the dual process theories require a mechanism to explain how the mind detects when analytical reasoning is necessary: what makes us think (Pennycook 2017).

As highlighted, dual process theories lack of solid explanation of how both perceptual and analytical thinking (or reasoning) are actually triggered. This challenges our true understanding of non-standard cognitive architecture and naturally leads to the crucial question of how to make the assumptions underlying a descriptive economic theory more accurate.

A New Perspective to Support the Notion of Non-standard Rationality in Real-World Simulation

For sure behavioural economics contributed to introduce new parameters in the utility function notion so to account for psychological factors ignored by the standard view, but if behavioural economic theories are still built in the same way as standard economic theories, they are likely to face the same methodological problems. The so called de-idealization of the homo economicus is probably not sufficient to account for the cognitive complexity of human decision-making: behavioural economics requires additional theories explaining the value of the parameters introduced and eventually a more radical change in its theoretical framework. In view to develop an economic theory able to model critical issues of concrete reality, it's probably necessary to give up mathematical modelling techniques and appearance of completely value-free objectivity.

As clearly described, the new perspective concerning individual non-standard rationality attitudes justify assumptions concerning two reciprocally influencing levels of cognition, within which beliefs and preferences are respectively encoded according to different kinds of representations (conscious and unconscious) and subject to different kinds of reasoning (automatic/unreflective and deliberative/reflective). Basing on our previous considerations, there is the need of a theory about preference formation (towards sensory or abstract state of affairs) in view to account for an adequate theory of individuals' non-standard decision-making attitudes. In fact, besides the possible justification of unconscious beliefs cognition and unconscious cognitive reasoning, there is the need to shed light on the unconscious mechanisms of preferences. Because preferences are responsible for all kinds of motivational attitudes towards decision-making, they trigger the attraction towards both normative and descriptive representations of objects, whatever the properties they feature (sensory or abstract); e.g. they are responsible for both consuming options and for the conformity to social/normative pattern of action. It's important to consider how any new contributions to a theory of preferences inevitably lead to changes in the epistemological and ontological representation of individual cognitive perspectives.

Challenging the Epistemological Paradigm of Current Decision-Making Models

The mainstream argument in standard rational choice epistemology was the maximization principle. After having tested the psychological impossibility that such kind of consistency could feature invariance properties in individuals' rational decision-making, the notion of analytical rationality has become an uncertain benchmark of correctness.

Changes in the epistemological perspective are mainly due to a new reference for consistency: the standard paradigm conceived the epistemological consistency as the rational adequacy to the maximization principle, so that the formal epistemic benchmark was a "rational" utility function. The new perspective suggests a new kind of epistemological consistency, namely, a consistency linked to functions of preferences, which can be variously encoded at different levels of individual cognitive attitudes; therefore the epistemological benchmark can be one of the possible preference functions encoded also at unconscious/sub-personal level. In this frame, we can consider a shift from an epistemological perspective supported by the concept of rational consistency to an epistemological perspective supported by the concept of logical consistency. The main difference between the two types of consistencies is that while rational consistency is supported by analytical/deductive reasoning, logical consistency can be supported by default (functional) reasoning, which is notoriously more rapid, automatic and not necessarily subject to deliberative thinking.

So individuals' intentions to choose or to act in a certain direction, which appear as deviations from the expected utility principle (standard epistemological benchmark), are not to be considered as paradoxes (Marquès and Weisman 2016), but rather as the expression of a different epistemological consistency of individuals' intentions. In this frame, it's important to consider how individual intentional attitudes can be supported (even at the same time) by two different types of consistent thought: one, which is consistent in rational-analytical terms, and another one, which is consistent in pure logical terms.

Importantly, a similar epistemological frame is linked to a cognitive perspective, which relies on two different but interactive levels of reasoning and also features a distinctive characteristic: complexity. As previously considered, preferences are not stable, rather can be strongly influenced by environmental cues; furthermore, they often work as functions of other relating (conscious or unconscious) preferences. This allows us to assume that systems of preferences are encoded as structures, which are everything but fixed and invariant; they can be rather dynamic systems, able to be influenced and even "reoriented" according to specific representations of context-dependent variables. Precisely, systems of preferences work as inner orders of fine-grained truth-functional relations, each representing a node of a motivational structure, which can generate new orders from time to time in a sort of inner equilibrium searching. As mentioned, it's plausible to assume that systems of preferences are progressively built at both conscious (reflective) and unconscious (unreflective) levels of cognition; this means that preference functions are encoded at each cognitive level. Precisely, each preference, respectively, anchors a structure of interrelated functional intentions, which then result in hierarchical as well as open (unreflective and reflective) systems. On the whole, systems of preferences will therefore correspond to systems of functional intentions, which work as cognitive vectors towards sensory and abstract objects of the surrounding environment (Pacherie 2006). This assumption is able to justify individual cognitive perspectives towards contexts, including salience of frames and social (normative) urgencies, which can exactly occur at both reflective and unreflective levels, according to the most dominant preference functions. Issues concerning social cognition and the normativity of social objects are especially important in this regard, but relating arguments cannot adequately be met in this occasion (Gerace 2012e, 2018).

It's sufficient here to consider that the non-standard rationality perspective entails a change in the epistemological paradigm, consisting in a shift of the benchmark of correctness and that the supporting cognitive view can assume systems of functional intentions, which express individuals' motivational attitudes towards objects of cognition, by constituting cognitive perspectives able to intercept salient properties in the surrounding environment.

What is then the justification for the "operational properties" of such cognitive systems? We may well accept the irrefutable analytical reasoning in support to the concept of rational preference, but how to account for the nature (and legitimacy) of the above-mentioned patterns of logical consistency concerning preferences individuals' mind?

Challenging the Ontological Paradigm of Current Decision-Making Models

We have seen how rational choice theory is accurate in showing how an ideal agent can rationally act, basing on his or her preferences, but how it fails in explaining where those preferences come from or how they might change. We have also seen how behavioural economics, in its accounts of non-standard rationality, lacks of an adequate theory of preferences, notwithstanding theoretical contributions from philosophy and from the social sciences and particularly historical contributions about the logic of preferences (Von Wright 1963) and about preferences change (Hansson 1995). A common distinction in economic theories and models is the one between exogenous and endogenous preferences, meaning the difference between preferences, which are established in advance as "rationally given" and are not explained by the model and preferences that cannot be taken as given but are generated within the model by means individuals' internal responses to decision-making environment. In standard economics, preferences are typically assumed to be fixed and exogenous: they constitute an essential feature of agents' identity, as well as it's not explained how they are formed. By contrast, behavioural economics reminds us that preferences are to be realistically assumed as endogenously generated by means of cognitive interactions between individuals' motivational attitudes and external information (Dietrich and List 2013). Specifically, according to behavioural economics, the ontological nature of preferences is partly "mentalistic" (preferences are mental states) and partly behaviouristic (preferences are concrete patterns of choice) (Heidl 2016). But such mentalistic approach is superficially assumed, only in view to explain "non-standard" behaviour and this doesn't do justice to any exhaustive theory of preferences.

As importantly considered (Dietrich and List 2013), the endogenous formation of preferences by means of cognitive interaction with the decision-making environment has not to be empirically intended as the processing of external information by means of an instrumental formal rationality but rather as the reception of information by cognitive perspectives based on "substantial rationality". While formal rationality is based on formal consistency with agents' beliefs and desires, the concept of substantial rationality implies not only formal consistency about agents' attitudes (preferences and beliefs) but also about their content (Dietrich and List 2013). Within a substantial rationality-based perspective, we can, for instance, consider the behaviour of an individual with self-destructive, or otherwise unreasonable preferences, not necessarily as irrational, but rather justify those preferences and resulting behaviour as internally consistent: the focus is on whether the agent's actions and choices are consistent with his or her preferences, not on whether those preferences are "formally reasonable". Although we can retrace typical features of substantive rationality in the neoclassical economic notion of self-interest, formal rationality and the relating possibility of probability (Bayesian) inferences are the basic concepts in economics, even in behavioural economics (Sugden 2005). By contrast, substantial rationality is a philosophical concept, which can be traced back to Kant's theoretical account of human reason as necessarily constituted by both contents and formal consistency (Kant 1781). Incorporating substantive accounts of rationality within a theory of individual preference (motivational attitudes) entails the need to account for the existence (or the constitution) of intrinsic practical "end" of the reason. For instance, according to self-interest theory, a familiar substantive theory of rationality, only options serving the end of self-interest maximization are to be preferred. This amounts in turn to the need of formalizing substantive accounts of rationality within a theory of individual preference by including normatively relevant properties of the alternatives/options of individual choice, in order for such options to be cognitively salient (or not) (Sudgen 2005; Dietrich and List 2013): in fact, according to self-interest theory, the only normative relevant properties of an option are those that directly interest the agent. Normativity in this regard is the concept able to express both the automatic preferability (on the side of the object) and the active choice (on the side of the agent) of a given option. By referring to a specific end, the normativity of a given option directly "affects" individual willingness in terms of (reflective or unreflective) motivational attitudes, exactly because that end has been (necessarily) constituted by willingness itself. With reference to a theory of preferences based on a dual system model of cognition, this amounts to the assumption that internal acts of willingness can sanction possible ends (i.e. preferences) as contents of possible substantial consistencies and that such acts of willingness can also perform at unreflective level of individual cognition. In this regard, substantial consistency is to be identified with the deriving functional systems of preferences, as responsible for cognitive perspectives, which are based on intentional/motivational attitudes and as such are able to intercept (or to percept), namely, to be affected, by certain salient normative properties of environmental contexts.

Our account of social normativity (Gerace 2012a, b, c, d, e, f, 2013a, b, c, 2018) and of the notion of material-based normative object (Gerace 2012a, c, d, e, f, 2013a, c) exactly retraces this perspective, which can be traced back to the cognitive notion of phenomenological essence (Gerace 2012b, d). One of the main tenets of this account is that substantial consistency is to be identified with an essential objectivity, namely, something, which rationally (or logically) self-determines and obtains independently from individuals' contingent attitudes. Therefore, in this perspective, any substantial consistency emerging as mental system (or sub-system) of preferences can be assigned the ontological status of object, precisely of normative object. Namely, each substantial consistency constitutes a specific "ontology" itself. Doxastic commitment is formally a primitive notion in this perspective: in each individual, from an original or contingent commitment, namely, from an act of willingness, able to sanction a certain preference for certain properties of specific environmental states of affairs can derive an internal system of functional preferences and relating intentional attitudes, to be intended as a normative object, elsewhere also identified with the notion of "value" (Gerace 2012a, b, c, d, e, f, 2013a, c). It's plausible to assume that systems of preferences can conflict at different cognitive levels. It's also a valid assumption to consider the unreflective/automatic motivational structure as able to be uncontrollably (emotionally) affected by knowledge of certain states of affairs or rather to unconditional support the direction of certain personal attitudes, such as action, choice and even deliberative (analytical) thinking. On the other side, it's plausible to assume a control role from the reflective/deliberative motivational structure, so to justify, for instance, the commitment behaviour of an individual fighting against his alcohol addiction, who may refrain from entering a bar because he thinks about the various tempting properties of wine that would become salient to him (Dietrich and List 2013).

Normativity can support the cognitive salience of whatever kind of object property, at both reflective and unreflective levels. It can even support the abstract properties of shared action patterns (social norms), at both heuristic and constitutive level (Gerace 2012a, b, c, d, e, f, 2013a, c), by means of derived functional preferences. In fact well-known psychological studies in social cognition (Cialdini et al. 1990) found evidence that people comply with social norms even in absence of any manifest incentive structure or direct personal commitment to what the norm stands for: assuming a conditional preference for following a fairness norm is different than assuming a fairness preference, i.e. the representational content of a social pattern is to be distinguished from the motivations supporting its compliance/enforcement. Unfortunately, arguments concerning the notion of normativity in social cognition cannot be adequately met in this occasion. Suffice it to consider, in this regard, the evidence that salience of social norms doesn't primarily derive from their representational content, rather from the (reflective or unreflective) motivational factors supporting them, which account for a purely normative cognition (Gerace 2018). If norms influence behaviour or decisional processes via representations of motivational and not descriptive factors, then the folk theory of social cognition is incomplete if not accounting for such representations.

On the whole, the advantages of an ontological paradigm of agents' decision-making, which is based on substantial normative consistency of choices, regard both the possibility to justify a theory of individual preference formation and change and the possibility to account for the real-world complexity of individual decisional attitudes.

Real-world simulations can keep track of systems of preferences as (material) content-based truth functional relations in individuals' mind (normative objects), which vary amongst people and societies. This is actually a specific perspective relating to the notion of value (Gerace 2012a, 2013c), to be intended as a cognitive notion, also in economic theory, for the sake of an adequate decision-making theory.

While the behavioural paradigm is supported by the idea that different contexts can give rise to different lenses through which the agent can acquire biased information and prefer (framed) options, the above-described perspective doesn't account for any individual personal evaluation or psychological "anomaly" in preference formation; rather preferences, as both information and motivational driven attitudes, represent themselves mind-dependent objectivities relating to the surrounding environment, to be legitimized and possibly traced back, at different cognitive levels.

So far models attempting to capture the role of perception in rational choice and, especially, to formalize the endogenous preference formation/change basing on individual motivational attitudes (Dietrich and List 2013) unfortunately continued to rely on a concept of instrumental practical rationality, without delving into the unreflective sphere of such motivational attitudes. As in standard rationality models, motivationally salient properties to choose x over y are interpreted as "reasons" and always based on agents' weighing relation over property combinations.

Knowledge Representation

As mentioned, systems of preferences are to be intended as open complex systems, since they can change (reorient themselves) and progressively feature emergent properties of the whole structure of motivational attitudes. In view to represent such preference systems, we don't want to build separate lists of object properties and of possible salient properties to be possibly matched (Dietrich and List 2013), because this would mean (again) to encapsulate the formation of preference within a set-up established in advance. We rather prefer to represent the substantial (material-based) truth-functional relations, which characterize the preference systems, as semantic validity nodes. The fine-grained ontology characterizing orders of fundamental and derived preferences can feature changes in the fundamental nodes, which can entail changes in the structure (derived preferences).

Importantly, the structures of preferences self-determine "objectively": having affirmed a material content or state of affairs "A", I am necessarily invited to accept a potentially infinite number of other state of affairs "$R_n(A)$" relating to the first one (in this sense the emergence and variation of preferences is linked to material contexts, able to orient individuals' perceiving attitudes):

$$\mathrm{A} \vDash \mathrm{non}\left(\mathrm{non\,A}\right)\mathrm{and}\left\{\mathrm{R}_1\left(\mathrm{A}\right),\mathrm{R}_2\left(\mathrm{A}\right),\ldots,\mathrm{R}_n\left(\mathrm{A}\right)_{,}\ldots\right\}.$$

The set of functional validities $\{R_1(A), R_2(A),\ldots, R_n(A)_{,} \ldots\}$ emerges as logical consequence and obtains according to an intrinsic consistency (where each validity term can be in turn represented as the origin of a new system, "founded" by the previous).

Depending on applications, the objects in the set can be material goods, outcomes of actions or social states of affairs: whatever object, able to be descripted in its properties, can represent a placeholder of semantic validity functions, which in the model stand for volitional (therefore normative) truth-functional relations.

Objects can be "placeholders" of functional relations. Namely, it's important to distinguish between the representation of descriptive properties and the representation of volitional relations. Substantial intentional or volitional truth-functional relations anchor descriptive representation of objects.

As understandable, an object "P" can be said to be salient and preferable if it falls within the scopes of an individual volitional (normative) perspective "p":

$$\mathrm{P} \leftrightarrow \exists x\,(\mathrm{pP}\left(\mathrm{x}\right)$$

The more the semantic validities of the normative perspective involved are founding, the more the salience of P is guaranteed.

The contents described in this paragraph derive from material elsewhere presented (Gerace 2012a, b, c, d, e, f, 2013a, b, c).

Concluding Remarks

We have described how the notion of non-standard rationality has basically inherited some conceptual premises of neoclassical economics. In this perspective, the supporting theoretical accounts hardly succeed in explaining exhaustive cognitive mechanisms for decision-making. Also in markets simulation, economists tend to follow the classical instrumental model of agency, whose weakness is its inability to account for preference formation or genuine preference change. Philosophers and social scientists, by contrast, tend to be interested in a more substantive model of agency, under which it's possible to account for the motivations that trigger the emergence and change of preferences. In this view, we have presented an alternative theoretical perspective, which expresses a substantive, not just formal, account of reasoning, able to extend the explanatory potential of the widely used belief-desire model of choice. The assumptions presented can elegantly support a theory of preference and possibly support simulation models of real-world decision-making, by relying on a refined as well as not intricate ontology.

References

Akerlof, G. A. (2002). Behavioural macroeconomics an macroeconomics behaviour. *The American Economic Review, 92*(3), 411–433.

Akerlof, G. A., & Kranton, R. E. (2000). Economics and identity. *The Quarterly Journal of Economics, 115*(3), 715–753.

Akerlof, G., & Shiller, R. (2009). *Animal Spirits: How human psychology drives the economy and why it matters for global capitalism*. Rinceton: Princeton University Press.

Allais, M. (1953). Le Comportement de l'Homme Rationnel devant le Risque: Critique des Postulats et Axiomes de l'Ecole Americaine. *Econometrica, 21*(4), 503–546.

Ariely, D. (2008). *Predictably irrational: The hidden forces that shape our decisions*. New York, NY: Harper Collins.

Ariely, D., Lowenstein, G., & Prelec, D. (2003). Coherent arbitrariness: Stable demand curves without stable preferences. *The Quarterly Journl of Economics, 118*, 73–105.

Baptista, M. L., et al. (2014). An agent-based model of consumer behaviour based on the BDI Architecture and Neoclassical theory. *Developments in Business Simulation and Experiential Learning, 41*, 170–178.

Bargh, J. A., & Morsella, E. (2008). The unconscious mind. *Perspectives on Psychological Science, 3*, 73–79.

Becker, S. G. (1976). *The economic approach to human behaviour*. Chicago: University of Chicago Press.

Bicchieri, C. (2006). *The grammar of society: The nature and dynamics of social norms*. Cambridge: University Press.

Binmore, K. (2010). Social norms or social preferences? *Mind and Society, 9*, 139–158.

Bratman, M. (1987). *Intentions, plans and practical reason*. Cambridge, MA: Harvard University Press.

Camerer, et al. (2003). Regulations for conservatives: Behavioural economics and the case for asymmetric paternalism. *University of Pennsylvania Law Review, 51*, 1211–1254.

Camerer, C. F., Loewenstein, G., & Rabin, M. (2004). *Advances in behavioral economics (The Roundtable Series in behavioral Economics)*. Princeton: Princeton University Press.

Cialdini, R. B., et al. (1990). A focus theory of normative conduct: A theoretical refinement and re-evaluation of the role of norms in human behaviour. *Advances in Experimental Social Psychology, 24*, 201–234.

Colombo, M. (2014). Caring, the emotions and social norm compliance. *Journal of Neuroscience, Psychology and Economics, 7*(1), 33–47.

Dahr, R., & Gorlin, M. (2013). A dual system framework to understand preference construction processes in choice. *Journal of Consumer Psychology, 23*(4), 528–542.

De Neys, W. (2014). Conflict detection, dual processes and logical intuition: Some clarifications. *Thinking and Reasoning, 20*(2), 169–187.

De Neys, W., & Glumicic, T. (2008). Conflict monitoring in dual process theories of thinking. *Cognition, 106*(3), 1248–1299.

DellaVigna, S. (2009). Psychology and economics: Evidence from the field. *Journal of Economic Literature, American Economic Association, 47*(2), 315–372.

Diasakos T. M. (2013). *Complexity and bounded rationality in individual decision problems*. DiscussionPaper series 201314, Department of Economics, University of St. Andrews.

Dietrich, F., & List, C. (2013). Where do preferences come from? *International Journal of Game Theory, Springer; Game Theory Society, 42*(3), 613–637.

Dijksterhuis, A., Bos, M. W., Nordgren, L. F., & van Baaren, R. B. (2006). On making the right choice: The deliberation-without-attention effect. *Science, 311*, 1005–1007.

Djiksterhuis, A., & Nordgren, L. F. (2006). A theory of unconscious thought. *Perspectives on Psychological Science, 1*, 95–106.

Elster, J. (1994). Rationality, emotions and social norms. *Synthese, 98*(1), 21–49.

Espinosa, B. A., et al. (2017). Toward formal modelling of affective agents in a BDI architecture. *ACM Transactions on Internet Technology, 17*(1), 5.

Evans, J. (2007). *Hypothetical thinking: Dual processes in reasoning and judgement*. New York: Psychology Press.

Evans, J. (2008). Dual-processing accounts of reasoning, judgment, and social cognition. *Annual Review of Psychology, 59*, 255–278.

Evans, J., & Stanovich, K. (2013). Dual-process theories of higher cognition: Advancing the debate. *Perspectives on Psychological Sciences, 8*(3), 223–241.

Fehr, E., & Gaetcher, S. (2000). Fairness and retaliation. The economics of reciprocity. *Journal of Economic Perspectives, 14*, 159–181.

Fisman, R., Shachar, K., & Markovits, D. (2007). Individual preferences for giving. *American Economic Review, 97*(5), 1858–1876.

Frank, R. H. (2006). *Microeconomics and behaviour*. McGraw-Hill/Irwin publisher.

Gerace, G. (2012a) *Justifying normative objects: An answer to J. Searle's model*. Paper presented at the 20th anniversary conference of the Italian Society for Analytic Philosophy (SIFA) – Sept. 12–15, 2012, Alghero (University of Sassari).

Gerace, G. (2012b) *Positionality and normative ought in Max Scheler's Formalism*. American Philosophical Association - Max Scheler Society of North America biannual conference (Chicago USA)

Gerace, G. (2012c). *Normative objects and material a-priori*. Poster presented at the conference "Objects in Mind" - Interacting Mind Centre, Aarhus University.

Gerace, G. (2012d). *Normativity and material-a priori*. Draft contents sent and discussed via e-mail with Francesco Guala; July the 7th 2012.

Gerace, G. (2012e). *On the Justification of normative objects*. Draft contents sent via e-mail to the attention of Mattia Gallotti and John Michael; December the 1st 2012.

Gerace, G. (2012f). *Justifying normative objects: An answer to J.Searle's model* Conference long abstract sent and discussed with Enrico Terrone; September the 18th 2012.

Gerace, G. (2013a). *Emergence and persistence of social norms: A material-based account*. Paper presented at the Mid-Term Postgraduate conference of the Italian Society of Logic and Philosophy of Science –SILFS - May 29–31, 2013, University of Urbino.

Gerace, G. (2013b). *Coordinate action and individual intentions*. Poster presented at the conference "Minds in Common" Institut d'Etude de la Cognition, École normale supérieure.

Gerace, G. (2013c). *Justifying normative objects: A material apriori-based account*. Phd Project Proposal submitted at the University of Exeter - College of Social Sciences and International Studies.

Gerace, G. (2018). *A study for the representation of social normativity*. Phd Project Proposal submitted at the IMT – Lucca and sent to the attention of Gustavo Cevolani; August the 3rd 2018.

Gigerenzer, G. (2008). Why heuristic works. *Perspectives on Psychological Science, 3*, 20–281.

Gigerenzer, G., & Goldstein, D. (1996). Reasoning the fast and frugal way: Models of bounded rationality. *Cognitive Science, 103*(4), 650–666.

Goodwin, N., et al. (2014). *Microeconomics in context*. New York: Routledge. Gordon, R.

Haidt, J. (2001). The emotional dog and its rational tail: A social intuitionist approach to moral judgment. *Psychological Review, 108*(4), 814–834.

Hansson, O. (1995). Changes in preferences. *Theory and Decision, 38*(1), 1–28.

Hassin, R., Uleman, J. S., & Bargh, J. A. (Eds.). (2005). *The new Unconscious*. Oxford, UK: Oxford University Press.

Hausman, D. (2012). *Preference, value, choice and welfare*. Cambridge: CUP.

Hédoin, C. (2015). From utilitarianism to paternalism: When Behavioural economics meets moral philosophy. *Revue de Philosophieèconomique, 16*, 73–106.

Heidl, S. (2016). *Philosophical problems of Behavioural economics*. London: Routledge.

Heukelom, F., & Sent, E.-M. (2017). Behavioral economics: From advising organizations to nudging individuals. *Journal of Behavioral Economics for Policy, 1*(1), 5–10.

Hirshleifer, D., & Shumway, T. (2003). Good day sunshine: Stock returns and the weather. *Journal of Finance, 58*(3), 1009–1032.

Infante, G., Lecouteux, G., & Sugden, R. (2016). Preference purification and the inner rational agent: A critique of the conventional wisdom of behavioural welfare economics. *Journal of Economic Methodology, 23*(1), 1–25.

Kahneman, D. (2003). Maps of bounded rationality: Psychology for behavioral economics. *American Economic Review, 93*(5), 1449–1475.

Kahneman, D. (2011). *Thinking, fast and slow*. London: Allen Lane.

Kahneman, D., & Frederick, S. (2005). A model of heuristic judgment. In K. Holyoak & R. G. Morrison (Eds.), *The Cambridge handbook of thinking and reasoning* (pp. 267–294). Cambridge, UK: Cambridge University Press.

Kahneman, D., & Tversky, A. (1979). Prospect theory: An analysis of decision under risk. *Econometrica, 47*(2), 263–292.

Kahneman, D., Knetsch, J., & Thaler, R. H. (1990). Experimental tests of the endowment effect and the Coase theorem. *Journal of Political Economy, 98*(6), 1325–1348.

Kant, I. (1781). Kritik der reinen Vernunft. De Gruyter Verlag, 2018.

Levin, J., & Milgrom, P. (2004). *Introduction to choice theory*. [online] pp. 1–25. Available at: https://web.stanford.edu/~jdlevin/Econ%20202/Choice%20Theory.pdf.

Lieberman, M. D. (2003). Reflective and reflexive judgment processes: A social cognitive neuroscience approach. In Forgas et al. (Eds.), *Social judgment: Implicit and explicit processes*. New York, NY: Cambridge University Press.

Logan, G. D., & Crump, M. J. (2010). Cognitive illusions of authorship reveal hierarchical error detection in skilled typists. *Science, 330*, 683–686.

Lowenstein, G., & Lerner, J. (2003). The role of affect in decision making. In R. Dawson, K. R. Scherer, & H. H. Goldsmith (Eds.), *Handbook of affective science* (pp. 619–642). Oxford: OUP.

Marquès, G., & Weisman, D. (2016). Rethinking Rationality theory's epistemological status. Normative vs positive approaches. *Real World Economic Review, 77*, 50–63.

Marshall, A. (1890). *Principles of economics*. London: Macmillan and Company.

Martin, J. W., & Sloman, S. A. (2013). Refining the dual-system theory of choice. *Journal of Consumer Psychology, 23*(4), 552–555.

Mathis, K., & Steffen, A. D. (2015). From rational choice to behavioural economics. In K. Mathis & A. Tor (Eds.), *Nudging - possibilities, limitations and applications in European law and economics*. Springer.

Melnikoff, D. E., & Bargh, J. A. (2018). The mythical number two. *Trends in Cognitive Sciences, 22*(4), 280–293.

Nickerson, R. S. (1998). Confirmation bias: A ubiquitous phenomenon in many guises. *Review of General Psychology, 2*, 175–220.

Pacherie, E. (2006). Towards a dynamic theory of intentions. In S. Pockett, W. P. Banks, & S. Gallagher (Eds.), *Does consciousness cause behavior? An investigation of the nature of volition* (pp. 145–167). Cambridge, MA: MIT Press.

Page, S. (2012). Aggregation in agent-based models of economics. *The Knowledge Engineering Review, 27*(2), 151–162.

Paternotte, C., & Grose, J. (2012). Social norms and game theory: Harmony or discord? *British Journal for the Philosophy of Science, 64*, 551–587.

Pennycook, G. (2017). A perspective on the theoretical foundation of dual process models. In W. De Neys (Ed.), *Dual process theory 2.0*. New York: Psychology Press.

Rabin, M. (1993). Incorporating fairness into game theory and economics. *The American Economic Review, 83*(5), 1281–1302.

Rabin, M. (2000). Risk aversion and expected utility theory: A calibration theorem. *Econometrica, 68*, 1281–1290.

Rabin, M. (2002a). A perspective on psychology and economics. *The European Economic Review, 46*(4–5), 657–685.

Rabin, M. (2002b). Inference by believers in the law of small numbers. *Quarterly Journal of Economics, 117*(3), 775–816.

Rabin, M., & Thaler, R. (2001). Anomalies: Risk Aversion. *Journal of Economic Perspectives, 15*, 219–232.

Rizzi, J. (2009). *Behavioral basis of the financial crisis*. Senior Investment Strategist, CapGen financial.

Rumbell, T., et al. (2012). Emotions in autonomous agents: Comparative analysis of mechanisms and functions. *Autonomous Agents and Multi-Agent System, 25*(1), 1–45.

Samson, A. (2016). *The behavioral economics guide*. London: Behavioral Science Solutions Ltd.

Samuelson, P. A. (1948). *Economics, an introductory analysis*. New York: McGraw-Hill Company.

Secchi, D., & Neumann, M. (Eds.). (2016). *Agent-based simulation of organizational behavior: New Frontiers of social science research*. New York: Springer.

Shane, F., Loewenstein, G., & O'Donoghue, T. (2002). Time discounting and time preference: A critical review. *Journal of Economic Literature, 40*(2), 351–401.

Sharpanskykh, A. (2007). Agent-based modeling of human organizations. In F. Amblard (Ed.), *Proceedings of the 4th European social simulation association conference* (pp. 335–347).

Simon, H. A. (1982). *Models of bounded rationality: Empirically grounded economic reason*. Cambridge, MA: MIT Press Ltd.

Simon, A., & Voyer, B. (2014). Emergency purchasing situations: Implications for consumer decision-making. *Journal of Economic Psychology, 44*, 21–33.

Slovic, P., et al. (2007). The affect heuristic. *European Journal of Operational Research, 177*(3), 1333–1352.

Smith, A. (1759). *The theory of moral sentiments. (1986)*. Düsseldorf: Wirtschaft und Finanzen.

Stanovich, K. E., et al. (2011). The complexity of developmental predictions from dual process models. *Developmental Review, 31*, 103–108.
Sugden, R. (2005). Why rationality is not a consequence of Hume's theory of choice. *European Journal of the History of Economic Thought, 12*(1), 113–118.
Chai Sun-Ki. (2001). *Choosing an identity: A general model of preference and belief formation*. Ann Arbor: University of Michigan Press.
Sunstein, C. R., & Thaler, R. H. (2003). Libertarian paternalism is not an oxymoron. *The University of Chicago Law Review, 70*(4), 1159–1202.
Tesfatsion, L. (2002). *Agent-based computational economics*. Growing economies from the bottom-up. *Artificial Life, 8*(1), 55–82.
Thaler, R. H. (1985). Mental accounting and consumer choice. *Marketing Science, 4*(3), 199–214.
Thaler, R. H. (1999). Mental accounting matters. *Journal of Behavioral Decision Making., 12*(3), 183–206.
Thaler, R. H. (2016). Behavioural economics: Past, present and future. *American Economic Review. American Economic Association, 106*(7), 1577–1600.
Thaler, R. H., & Sunstein, C. R. (2008). *Nudge: Improving decisions about health, wealth, and happiness*. New Haven: Yale University Press.
Tversky, A., & Kahneman, D. (1974). Judgment under uncertainty: Heuristics and biases. *Science, 185*(4157), 1124–1131.
Velmans, M. (2000). *Understanding consciousness*. London: Routledge.
Von Wright, G. H. (1963). *The logic of preference*. Edimburgh: Edimburgh University Press.
Weber, R., & Dawes, R. (2010). Behavioural economics. In N. J. Smelser & R. Swedberg (Eds.), *The handbook of economic sociology*. Princeton: Princeton University Press.
Windrum, P., et al. (2007). Empirical validation of agent-based models: Alternatives and prospects. *Journal of Artificial Societies and Social Simulation, 10*(2), 1–8.

Chapter 4
The Many Faces of Crowdfunding: A Brief Classification of the Systems and a Snapshot of Kickstarter

Marco Campennì, Marco Benedetti, and Federico Cecconi

Abstract In this chapter, we present the complex phenomenon of crowdfunding and its origin, briefly providing some useful definitions and concepts. We then present two different types of analysis: (i) a simple descriptive statistics showing interesting results about the most successful categories of Kickstarter campaigns; (ii) the potentials of adopting a time-series analysis approach to better understand trends in Kickstarter and to perform a simple forecast of future trends of successful campaigns.

Keywords Crowdfunding · Kickstarter · Time-series analysis · Reward-based · Multi-actor system

Introduction

In this chapter, we present the complex phenomenon of crowdfunding and its origin, briefly providing some useful definitions and concepts.

We then present two different types of analysis: a simple descriptive statistics showing interesting results about the most successful categories of Kickstarter campaigns; the potentials of adopting a time-series analysis approach to better understand trends in Kickstarter and to perform a simple forecast of future trends of successful campaigns.

M. Campennì (✉)
Biosciences, University of Exeter, Penryn, Cornwall, UK
e-mail: m.campenni@exeter.ac.uk

M. Benedetti
University of Rome Sapienza, Rome, Italy

F. Cecconi
LABSS ISTC-CNR, ROME, Italy

© Springer Nature Switzerland AG 2019
F. Cecconi, M. Campennì (eds.), *Information and Communication Technologies (ICT) in Economic Modeling*, Computational Social Sciences,
https://doi.org/10.1007/978-3-030-22605-3_4

Definition and Origins of Crowdfunding

The term crowdfunding derives from the English crowd and funding, a process of collaboration between several people who decide to allocate their money to support the efforts, projects and visions of other private citizens, but also of companies, institutions and organizations. It represents a relatively new method to obtain funding, and it works leveraging on the potential massive audience of Internet users.

Crowdfunding can be defined as an open call for resources implemented by project promoters via Internet aiming to receive donations or funds in exchange of a reward or the right of having access at the product(s) before the final commercialization or at a reduced price. This flexible definition highlights the original structure of this financial method which is able to adapt itself to very different situations depending on the promoted product and the targeted audience. This is why there are different forms of crowdfunding varying on the purpose of the campaign, on the type of reward and on the scale of the expected implementation.

A fundamental element for the success of the campaign is the emotional involvement of the crowdfunders. They participate in order to feel part of a community.

Irrespective of the specific purpose of the promoted campaign or the needs of the potential backers, it is always possible to identify three main components involved in the whole process: the campaign promoters, the backers and the (online) platform used to promote the campaign and to collect funds.

- The campaign promoters are the reason for the existence of the crowdfunding itself: they want to promote their ideas and projects, they need a platform to advertise them, and most importantly they need backers willing to pay for a form of reward.
- Backers represent the potential campaign promoter's ideas and projects. They (micro-)finance the campaigns for different reasons with the expectation (or not) of a reward and depending on different factors, such as their interests, needs or simply willingness of helping others donating an amount in order to help someone else in implementing a campaign. Usually, at least this is what campaign promoters aim to, backers are numerous and worldwide ubiquitous and invest a small amount of money and potentially invest in more than a single campaign.
- Finally, the online platform: it allows the connection between demand and supply. It plays the role of the (virtual) space where promoters and backers can meet and "play the crowdfunding game" following the rules both sides accepted to comply with.

The very first form of crowdfunding we can historically identify is from the beginning of the eighteenth century, when Jonathan Swift created the "Irish Loan Fund", a proto form of micro-credit. This fund allowed families with low income to benefit from small loan without providing any financial guarantee.

More than two centuries after in Bangladesh, Dr. Muhammad Yunus established a financial project allowing people with low income to start their own business activity. In few years, the project has exponentially grown and eventually became the Grameen Bank.

Finally, between the early 1990s and early years of the twenty-first century, the advent and the outbreak of the online sharing platforms changed dramatically the landscape of (micro)economic transactions and de facto allowed the crowdfunding to invade the cyberspace.

Particularly, between 2008 and 2009, two of the most well-known crowdfunding platforms started regulating micro-financial transactions, Kickstarter and IndieGoGo. Their characteristics may be summarized into the idea of promoting the creativity of small entrepreneurs allowing them to reach (at least) the production phase of their project (assuming, of course, that they are able to reach the goal of the campaign they are promoting). Moreover, the advent of online crowdfunding gave birth to the new idea of receiving different types of reward in exchange of the financial support to a promoted campaign.

One of the most recent and crucial phases of the online crowdfunding development is represented by the Jumpstart Our Business Startups (JOBS) Act, a law intended to encourage funding of small businesses in the United States by easing many of the country's security regulations. It passed with bipartisan support and was signed into law by President Barack Obama on April 5, 2012.

Why and How Crowdfunding Developed so Much?

Different factors allowed crowdfunding to emerge as a successful financial model. We can easily identify at least three concurrent phenomena:

The scarcity of resources at financial and capital level brought by the recent economic crisis, which has made it terribly difficult for companies, especially in the initial phase of their activity, to find funds from the main financial intermediaries. More specifically, we are talking about the global crisis that first broke out in the United States in 2007 and subsequently in the rest of the world that brought with it a series of financial resource limitations and access to credit. The establishment of a much more austere and severe regime for obtaining funds from classic financial intermediaries, such as banks, has denied the possibility of developing and creating entrepreneurial initiatives as well as other non-economic projects. This negative trend has found, therefore, a response in crowdfunding, defined as an alternative method of financing since, in fact, it does not take into consideration the main sources of financing (banks, Venture Capital companies, Business Angels). The request for funds from a very large audience of people has thus allowed to bypass, at least theoretically, the recent economic crisis with the consequences on the efficiency of the financial markets that it has brought with it.

The second factor is crowdsourcing. The term derives from the juxtaposition of two words: crowd and outsourcing, the practice of entrusting certain business activities to third parties. The credit for introducing the term goes to Jeff Howe and Mark Robinson, who used it for the first time in Wired Magazine in 2006. Crowdfunding, as previously mentioned, is among the sub-categories of crowdsourcing, presenting differences and similarities with it. The fundamental concept on which to emphasize the definition is the presence of the "Internet crowd" with the aim of offering a "voluntary contribution" to the business cause; in the case of crowdfunding, finding the necessary funds to implement an entrepreneurial initiative. With the advent of crowdsourcing, therefore, there has been the variation of its various meanings.

The evolution of Web 2.0 and related technologies created a direct and faster channel for an effective interaction between the various categories of Internet subjects. Without the Web 2.0, the possibility for companies to receive and forward information and requests to and from the public would have been an overly arduous task. From the technological point of view, Web 2.0 refers to the possibility given to computers to automatically process information, thus being able to spread and recombine them more easily. From a sociological point of view, however, it is the basis for creating a network of people who share common interests, and it is many orders of magnitude bigger than the social network of relatives, friends and colleagues. Finally, from an economic point of view, Web 2.0 is characterized by the possibility given to everyone to create content and put it into "the network". The incentives to create these contents derive from expectations of future rewards, in material or monetary form as well as purely intangible (e.g. mere gratification). Moreover, the advent of the Web 2.0 contributed to the emergence and stabilization of crowdfunding with additional factors such as the improved opportunities for collaborations, with the possibility of sharing everyone's resources and knowledge, the possibility of contributing freely to the various projects and the easy access and use of computers and the Internet, with the result of increasing user participation.

We can identify four main crowdfunding models:

- *Donation-based*
- *Reward-based*
- *Lending-based*
- *Equity-based*

The aforementioned classification is based on the type of exchange that is created between the proposer, also called creator (or campaign promoter), or the one who presents the idea and then obtains financing and the lending subject, called crowdfunder (or backer). Irrespectively from the economic contribution, the reward for the support can be of a material or intangible nature; however, the relationship that is created between the two subjects is not necessarily based on an exchange, in fact there are campaigns related to the donation-based model or providing a non-monetary reward.

Differentiated Support: Donations, Active Investments and Passive Investments

We can make a distinction between donations, investments with active participation of the crowd and, finally, passive investments.

The first variant, the donations, concerns fundraising projects that are based on the voluntary contribution of online supporters without a reward of any kind in return.

Active investments are characterized by the coexistence of a reward for the investors, financial or otherwise, and their active participation in the cause of the project. This reward and participation scheme can have positive repercussions on the management of the business project and on its good implementation, favouring feedback and advice to the entrepreneur. Through participation in the business profit and the acquisition of voting rights, the crowd of investors manages to influence entrepreneurial actions by offering their own vision and support; this regards not only the characteristics inherent to the product or service but also with reference to one's expectations about the demand and the potentials of the market. This is what usually happens in the crowdsourcing phenomenon.

The last category, that of passive investors, presents a multitude of financiers without any management power or right to respect to the financed company. The entrepreneur, in this case, prefers to make use of the crowd only to obtain funds, without losing a minimum of control or profiting from external ideas.

Donation-Based Crowdfunding, Equity-Based Crowdfunding, Reward-Based Crowdfunding and Crowdlending

Let us now analyse the four main existing business models for crowdfunding.

Donation-based crowdfunding: this type of fundraising is characterized by the absence of a form of reward for the backers. The reasons why individuals decide to "donate" part of their funds can be explained as "philanthropic" participation in supporting crowdfunding projects. Backers in this perspective finance projects despite the absence of a financial reward or not, for personal, social and, indeed, philanthropic reasons.

Equity-based crowdfunding: in this model, the backers of the entrepreneurial idea opt for a reward in the form of shares of the new company in favour of the financier individuals. An entrepreneur sells a certain amount of shares of the company to a group of (small) investors through an open invitation to obtain financing on an Internet platform. With equity-based crowdfunding, the ownership of the new activity will be divided among a multitude of small (and sometimes very large) financiers, with relative loss of control of the promoters. The reward, however, is indirectly donated with the shareholding, being, instead, directly correlated with the future performance of the financed company and therefore with the possibility of enjoying dividends or capital gains in the event of sale of the share.

Reward-based crowdfunding: a different method of crowdfunding consists in the financial support in exchange of a non-monetary reward. This reward can be of different forms, allowing, for instance, backers to have preferential access to the promoted product before the commercialization. More specifically, we can identify four main types of reward on the popular Kickstarter platform, including a copy of the product, a creative collaboration of various kinds, creative experiences pertaining to the object of the company and, finally, an imaginative "souvenir" for financial help.

Crowdlending: the fourth type of crowdfunding is also called peer-to-peer loan and is a direct alternative to the bank loan. The scheme is simple, with a contribution in cash giving rise to a certain interest rate and specific deadlines. The difference with a bank loan lies in the number of individuals from whom the loans are obtained, being in number concretely greater in crowdlending than in the usual single source in the case of banking transactions. The platform is used to bring together potential investors with the subject in financial deficit; the promoter of the campaign is subsequently legally bound to the repayment of the borrowed funds, becoming the debtor of multiple small creditors.

Advantages and Risks of the Crowdfunding System

The positive features of using the Internet crowd to obtain financial resources lie mainly in non-monetary factors.

A first fundamental benefit offered by crowdfunding is represented by having confirmation from the web population and potential consumers of the validity and acceptance of the idea. Supporting the idea by offering money or simple interest in crowdfunding campaigns, in exchange for various kinds of rewards, confirms the public's interest and the existence of a potential market for the product.

A successful crowdfunding campaign can confirm the existence of a market for the proposed idea and therefore the presence of potential consumers. This demonstrates the goodness of the project and is the first crucial step for entrepreneurs to get other sources of financing by traditional investors (e.g. banks or big companies).

Compared to traditional fundraising methods, crowdfunding offers advantages in terms of cost reduction and speeding up the development and support procedures of the project.

One of the peculiarities of crowdfunding is the number of potential supporters of the idea. Speaking to the Internet audience, an individual, in exploiting the potential of crowdfunding, can use the opportunity to approach and be approached by a particularly large number of subjects. This possibility allows an interaction and an exchange of ideas, opinions and feedback between the "backers" and the "promoters".

Another fundamental advantage is using the online platform for marketing activities. The dynamic and intelligent use of the platform, in fact, can contribute to spread certain messages or contents with very specific purposes, in order to create engagement or interest in the idea and thus increase the probability of success.

But, crowdfunding can also be very risky for both backers and campaign promoters.

One of the risks is represented by the fact that it is impossible being sure of the success of the project, with the undesired effect of paying for something that will never see the production phase.

Another inconvenience is that investors use the decisions of others as an informative signal on the quality of the project. So, in the event that the campaign finds little initial success or complaints from a certain number of investors, there will be negative repercussions on the probability of success of the campaign.

Another risk is usually defined as "adverse selection bias", and it refers to the idea that potential investors might have a bad interpretation of the choice of using crowdfunding, suggesting that the promoters of the campaign were not able to find funds in a more traditional way and therefore that, ultimately, the quality of the project is poor.

Finally, there is the risk represented by sharing information and sensitive details of the promoter's idea (i.e. the intellectual property issue). This may happen depending on the rules of the platform used to promote the idea and usually is caused by the need of information transparency. Without the exposure of specific and relevant details of one's idea, the information asymmetry between backers and promoters would be such as to prevent the former from carrying out an in-depth analysis of the entrepreneurial project, causing them to avoid investments.

Another relevant risk is represented by the exploitation of crowdfunding with criminal intents. Crowdfunding platforms can be used to promote fake campaigns with the intent of robbing investor funds. The phenomenon is favoured by the small amounts invested in the various campaigns, which de facto becomes an insufficient stimulus to the adoption of legal methods in response to fraud.

Moreover, the incompetence of the entrepreneurs promoting the campaigns can produce exactly the same effect. An overestimation of the potential of an idea or its incorrect implementation can lead to the failure of the project and the loss of the investment as a whole. To overcome this problem, the collective behaviour of the crowd may help potential investors in properly evaluating the quality of the project itself; if discouraging signals arrive from the crowd about the idea or if the project has not collected enough, a potential funder may be discouraged to invest.

Kickstarter: A Snapshot of the Current State

We have analysed the behaviour of campaigns of Kickstarter using the log available at the address: https://webrobots.io/kickstarter-datasets/.

The complete dataset has 207,848 records for all campaigns from January 2019; we used only data from successful and unsuccessful campaigns, respectively, 118,175 and 75,481 records (overall, 193,656 records).

To each campaign, a different category is assigned for an overall number of 169 different categories.

With successful campaigns, we refer to campaigns reaching at least the amount set as the "goal" of the campaign, and with unsuccessful (or failed) campaigns, we refer to campaign that failed in reaching the "goal". In the original dataset we used to run this analysis, there are also other campaigns with a different state (see Table 4.1 for details about frequencies of different state).

Figure 4.1 shows the distribution of the success rate of successful campaigns as per the month of January 2019.

Success rate is calculated as the ratio between amount reached at the end of the campaign and the goal set before the campaign starts. Values expressed in the dataset in US dollars.

Figure 4.1 shows how the distribution of the mean value of success rate of campaigns by category is not normally distributed but rather follows a power law, where a large number of campaigns grouped by category have a low success rate and only a few of them have a high success rate. The same type of distribution is obtained by calculating the frequency distribution of which the categories are most likely to succeed (Fig. 4.2).

It would therefore seem very important which campaign is proposed, to be able to predict its progress. Most successful campaigns come from well-defined types. With a linear regression, we have estimated the categories that are the most relevant from the point of view of obtaining a successful result.

Table 4.2 shows the categories that are to be considered (with a significance of $p < 0.001$).

From the analysis, it is clear that the macro-category (or type) "technology" is one of the most successful with 6 occurrences of successful campaigns followed by design, film and video and games.

Kickstart 2016–2019 Data and Time-Series Analysis

Considering the data available on the Kickstarter databases, it is possible to observe the evolution of different aggregates over time. Technically, this practice is called time-series analysis. A time-series is a sequence of data points observed at regular time intervals. The time-series analysis is a tool used not only in economics and finance but also in medicine, biology and in any sector where it is possible to observe

Table 4.1 Frequencies of different state of campaigns as per January 2019

Cancelled	Failed	Live	Successful	Suspended
8681	75,481	4888	118,175	623

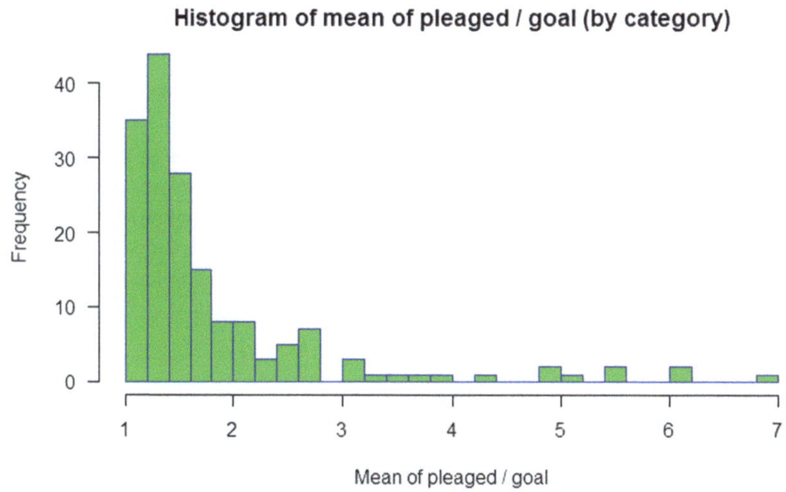

Fig. 4.1 Distribution of average success rate of successful campaigns by category as per January 2019

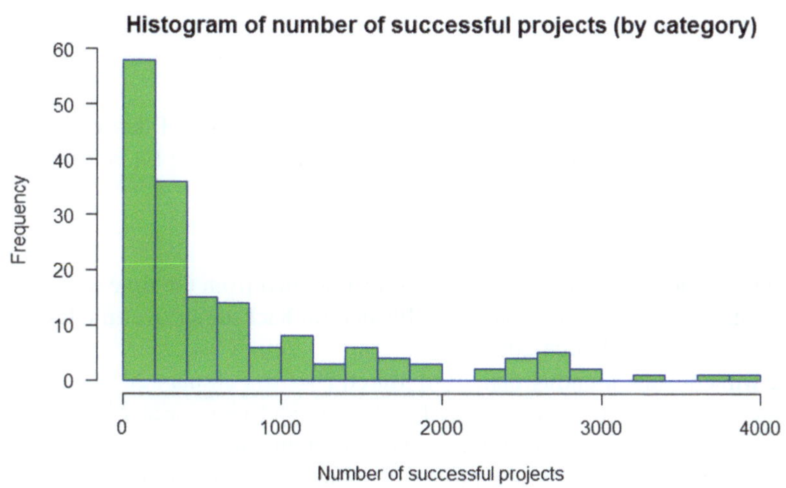

Fig. 4.2 Distribution of the number of success campaigns by category as per January 2019

temporal evolutions of certain processes. In the classical methodology, the objective is to decompose a historical series into its main components: the trend, the cycle, the seasonality and the random component. The trend represents the long-term trajectory and is one of the most interesting components to observe in case you want to predict future values for the aggregate under study. The cycle represents, especially in the case of economic time-series, the occurrence of contractions or expansions of the system, which cause oscillations around the trend. Seasonality, on the other hand, is given by fluctuations originating from climatic factors (alternation of

Table 4.2 This table shows the most successful categories of campaigns

Category	Type/category
Product design	Design/product design
Animation	Film and video/animation
3D printing	Technology/3D printing
Camera equipment	Technology/camera equipment
Gadgets	Technology/gadgets
Sound	Technology/sound
Wearables	Technology/wearables
Video games	Games/video games
Hardware	Technology/hardware

seasons) and/or social organization (holidays, etc.). The random component instead represents the residue of the model, i.e. the oscillations due only to an unpredictable component. The latter is the only stochastic component of the model.

These components can be combined through an additive or multiplicative relationship.

An additive model is when the fluctuations in the time series stay constant over time (see Eq. 4.1 for details).

$$Y_t = \text{Trend}_t + \text{Cycle}_t + \text{Seasonality}_t + \text{Noise}_t \tag{4.1}$$

A multiplicative model, instead, is when the fluctuations of the temporal series increase in the time and depends on the level of the series (see Eq. 4.2 for details).

$$Y_t = \text{Trend}_t \cdot \text{Cycle}_t \cdot \text{Seasonality}_t \cdot \text{Noise}_t \tag{4.2}$$

In this section, we will discuss the results obtained from the time-series analysis of the data regarding the campaigns published on Kickstarter over the last 3 years, from March 2016 to March 2019.

We will consider monthly data from March 2016 to February 2019. In particular, we will focus on the presence of any trends, considering separately the campaigns that have achieved their goal and those not. We define the success rate of any campaign as the ratio between the amount of money pledged and the goal set in US dollars (see Eq. 4.3).

$$\text{success.rate} = \frac{\text{amount.pledged}}{\text{amount.goal}} \tag{4.3}$$

The dataset used to run the analyses is accessible from: https://webrobots.io/kickstarter-datasets/.

We combined all files relative to months from March 2016 to March 2019 (not included), for a total amount of 6,135,833 records.

We only considered successful and failed campaigns, ignoring all campaigns with different state, such as cancelled, live and suspended.

We assumed that the data we considered can be analysed adopting an additive approach, assuming irrelevant the cyclic component and the seasonality independent from the trend (as it should be clear also form Fig. 4.1 about the decomposition of different components).

The success rate, for the part exceeding one, represents how much in percentage the goal has been overtaken (and therefore, the range of this success rate is between 0 and 1 or more, in case the Kickstarter campaign has been very successful receiving more funds than the goal). In order to take into account a single value for each month, the median is considered as representative statistic. This is the same with campaigns that do not achieve their goal (i.e. the "failure" data). If successful campaigns are considered (i.e. "success" data), on average they achieve about 19% more than their target. While for unsuccessful campaigns, on average they achieve only 1% of their goal. We can therefore observe that there is a large gap between the results obtained by successful campaigns and unsuccessful ones (see Fig. 4.3 for the plots of the two sets of data).

In order to evaluate the possible presence of a trend in the temporal evolution of the data about successful campaigns, we use the statistical methodology of the decomposition of time-series (see Fig. 4.4 for details).

This procedure is a statistical method that deconstructs a time series in several components: trend, seasonal, cyclical and irregular. We can assume the cyclical component to be irrelevant, because we are not considering a long time period of analysis. This analysis shows that the series of successes, namely, those campaigns that have achieved at least the set goal, presents an increasing trend, as can be seen from the graph in Fig. 4.4. This means that for those campaigns that reached the goal or performed even better, the amount collected in the future would be even greater.

Another important analysis we should run is about autocorrelation: this is the correlation of a time series with lags of itself. This is a significant metric because it shows whether the previous states (lagged observations) of the time series have an influence on the current state or not. In the autocorrelation chart, if the autocorrelation

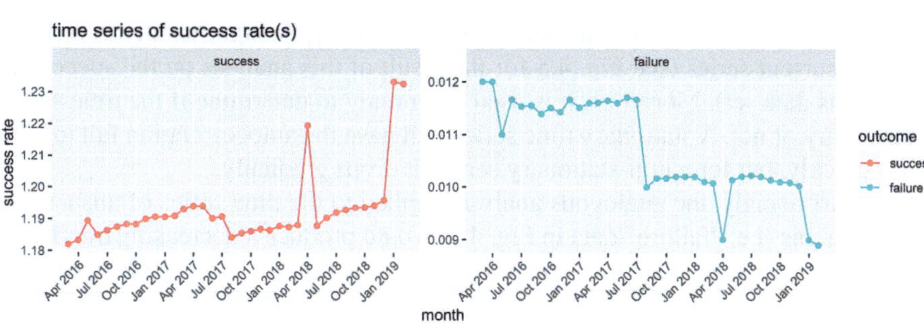

Fig. 4.3 Time series of successful and unsuccessful campaigns from March 2016 to March 2019 (not included)

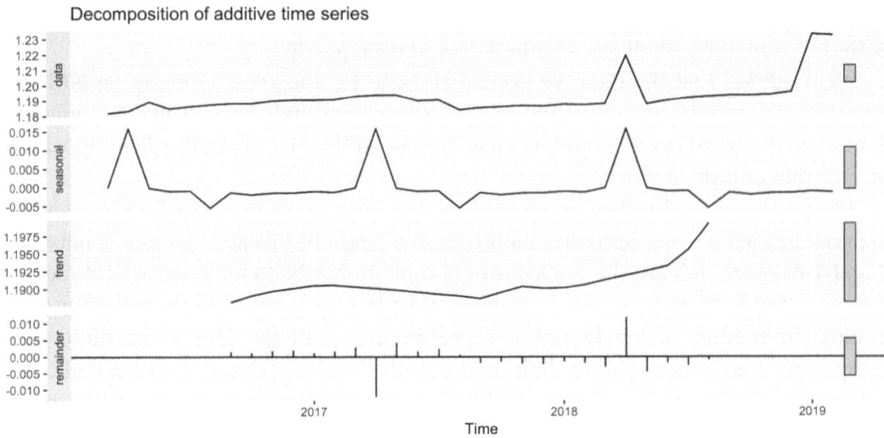

Fig. 4.4 Decomposition of the time series of successful campaigns

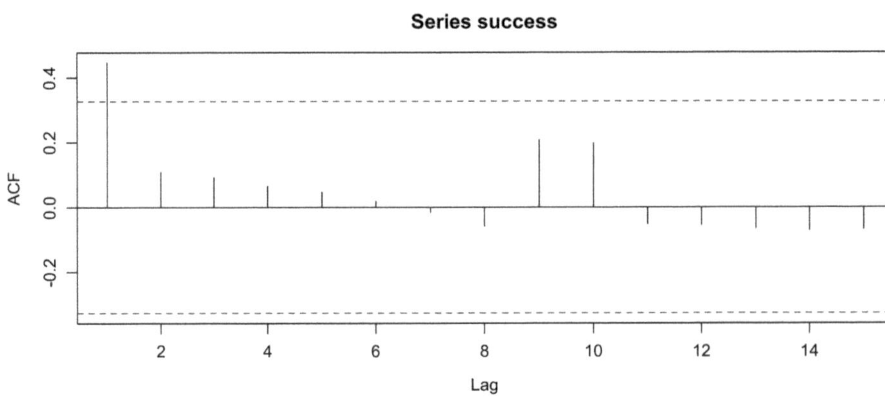

Fig. 4.5 This figure shows the autocorrelation function of the successful campaigns data set. The value of "lag" does not cross the blue line (except for the very first point, which is normal, being the first point of the series) nor collapses to zero

crosses the dashed blue line, it means that specific lag is significantly correlated with the current series (see Fig. 4.5 for the result of this analysis on the successful campaigns data set). Moreover, it is used commonly to determine if the time series is stationary or not. A stationary time series will have the autocorrelation fall to zero fairly quickly, but for a non-stationary series, it drops gradually.

Not surprisingly, the analogous analysis applied to the time series of unsuccessful campaigns (i.e. "failure" data in Fig. 4.3) would produce a decreasing trend, and since we are focusing our investigation on success rate of successful campaigns only and we are interested in showing how the method works and how it can be applied to any potential data, we will omit the analysis of unsuccessful campaigns, but the analysis we performed for successful campaigns could be easily applied to unsuccessful campaigns as well.

It is important to note that this result can only be read considering that the number of published campaigns, over time, has also been increasing. This means that not only has the number of successful campaigns increased in the last 3 years, but also the amount collected in relation to the goal set has increased. Once it is observed that there is a rising trend, it is possible to calculate a forecast of the future values of the series (see Fig. 4.6 for details). Given that the data used cover the last 3 years, the time horizon of the forecast is set at 4 months. This is because based on our experience we think that data available are not enough to perform a forecasting on a wider time-window.

With this purpose, we estimate the parameters of a linear regression by relating the percentage of gain to respect to the objective with time. The parameters of the regression line are both statistically significant, and in particular the angular coefficient is equal to 0.05%. This means that, with respect to this estimate, a monthly increase of 0.05% in the pledged/goal ratio is expected. As can be seen in the graph below, this analysis also confirms a growing trend for successful campaigns.

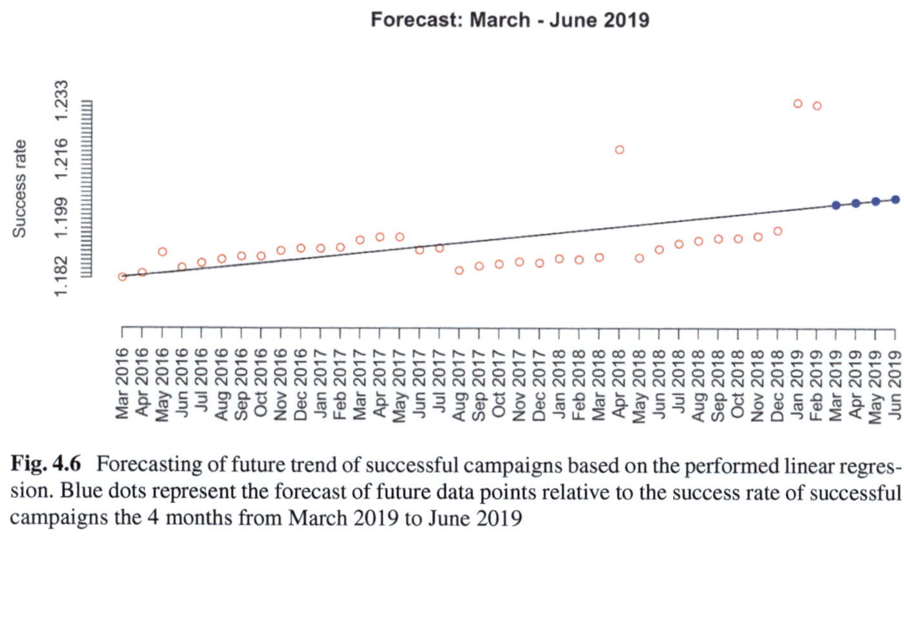

Fig. 4.6 Forecasting of future trend of successful campaigns based on the performed linear regression. Blue dots represent the forecast of future data points relative to the success rate of successful campaigns the 4 months from March 2019 to June 2019

Part II
Applications

Chapter 5
Passing-on in Cartel Damages Action: An Agent-Based Model

Claudia Nardone and Federico Cecconi

Abstract When there are infringements causing a price increase, such as cartels, one of the effects on direct purchaser's profit is the so-called passing-on effect. Passing-on rate is the proportion of the illegal price increase that cartel direct purchasers, in turn, translate into an increase in their own final price. In this chapter, we develop an agent-based model of a supply chain, where agents are firms who lay on different levels of the chain and are engaged in trading.

Keywords Price variation · Agent-based modelling · Trading · Firm behaviour · Cartel

Introduction

In the last few years, private actions for damages have been an important subject matter for discussion in the European antitrust rules' context. Infringements of the EU antitrust rules (Articles 101 and 102 TFEU), such as cartels or abuse of a dominant position in the market, are not only detrimental for the economy and consumers at large, but they also cause concrete harm, such as higher prices or lost profits, to concrete victims, e.g. infringers' direct and indirect customers, competitors and their customers. Therefore, at the end of 2014, the European Parliament and Council adopted Directive (EU) 2014/104, which introduces a common regulation for private claims for damages caused by infringements of competition law. In this context, quantification of antitrust damages plays a central role (European Commission 2013; Ashton and (Henry 2013). When there are infringements causing a price increase, such as cartels, one of the effects on direct purchaser's profit is the

C. Nardone (✉)
CEIS – University of Rome Tor Vergata, Rome, Italy
e-mail: claudia.nardone@uniroma2.it

F. Cecconi
LABSS ISTC-CNR, ROME, Italy

© Springer Nature Switzerland AG 2019
F. Cecconi, M. Campennì (eds.), *Information and Communication Technologies (ICT) in Economic Modeling*, Computational Social Sciences,
https://doi.org/10.1007/978-3-030-22605-3_5

so-called passing-on effect. A firm facing an input price increase probably will increase her own final price. Passing-on rate is the proportion of the illegal price increase that cartel direct purchasers, in turn, translate into an increase in their own final price (Verboven and Van Dijk 2009); Hellwig 2006). The extent of the costs translation into prices substantially varies from one case to another, because it strictly depends on a huge set of different factors, such as market structure, the degree of competition, buyer power, different prices strategies, etc. In order to estimate the actual pass-on rate, it is thus necessary to take into account all these aspects together. Therefore, quantification of the actual damage suffered by the victim requires the use of complex economic models and estimation techniques (Davis and Garcés 2009; Friederiszick and Röller 2010; Han etal. 2009; European Commission 2016).

An opportunity to study price variations, starting from direct interactions among firms, and considering different aspects of the "real world", without relying on stringent assumptions, is given by agent-based computational approach (Tesfatsion 2005; Tesfatsion and Judd 2006). We here develop an agent-based model to mimic trading between firms in a supply chain. Agents are firms who lay on different levels of the chain and are engaged in trading. At each level, firms buy the input from firms at the previous level and sell on the half-processed good to firms at the subsequent level. We are interested in what happens to prices when firms with capacity constraints compete both in price and quantity at the same time. We then introduce, at a certain production stage, a "cartel": some or all firms collude and set a price above the competitive level (Tirole 2003). In this way, we are able to quantify the pass-on rate. Theoretical (see for example Weyl and Fabinger (2009), Ashenfelter et al. (1998), RBB Economics (2014)) and empirical literature (see for example Kim and Cotterill (2008), Leibtag et al. (2007), Miller et al. (2016), Bonnet and Villas Boas (2013)) show that the extent of the cost translation into prices substantially varies from one setting to another, because it strictly depends on a huge set of different factors, such as market structure, the degree of competition, buyer power, dynamic changes in competition, different prices strategies, etc. To quantify the true pass-on rate, it is thus necessary to take into account all these aspects together. Here, we consider different numbers of firms involved in the illicit agreement and see how the pass-on rate changes in different scenarios. In this model, we try to solve some computational and behavioural problems in production chain pricing, not easily solvable within analytical frameworks, such as rationing processes, combined with the "minimum price" rule, and best responses to rationing processes.

The Agent-Based Model

Key Aspects

Here we present the main features of this model, following the pattern drawn in general for economic AB models presented in Chap. 1.

1. Heterogeneity. We consider a set of firms, the agents, hierarchically organized in a vertical supply chain, with several layers of intermediaries, each adding value to produce a homogenous consumer product. There is a number S of production stages (or layers) and a number n_s of active firms in stage s. Each firm has different costs and different capacity constraints which determines different choices of price and quantity offered, even if the good is homogeneous. This happens because of the supply capacity rationing that implies a certain degree of market power for firms.
2. Direct endogenous interactions. In our setting, each firm at any stage s purchases a homogeneous input from (one or more) firms in layer $(s - 1)$, input that she transforms using a one-to-one technology, into a homogeneous new output, which she sells on to (one or more) firms in layer $(s + 1)$. So, firms are engaged in trades, being both buyers and sellers. Firms who lay on the first upstream stage are supposed to buy from the outside world, while firms on layer S, retail layer, sell the final product to consumers, which are considered as an aggregate, and represented by a classic linear demand function. Agents interact directly, and the decisions undertaken today by an agent directly depend, through adaptive expectations, on the past choices made by other agents. In fact, firms set their own price and fix quantity offered, taking into account not only their own past performance but also the price decisions made by their competitors and the quantity offered by their supplier. In this way, the economy we describe is decentralized, and the resulting aggregate properties are not known a priori but obtained as the outcome of microdynamics, in line with the bottom-up perspective: agents are autonomous and no central or "top-down" control over their behaviour is admitted.
3. Bounded rationality. Bounded rationality arises because information is both private and limited and because agents are endowed with a finite computing capacity. In our model, each firm at a generic production stage s, during the period t, in addition to its past performance, is able to observe prices set by its competitors, i.e. firms who lay on the same stage s, and prices set by its suppliers, i.e. firms who lay on the previous stage $(s - 1)$. Firms do not know demand function nor variables affecting it: they can only observe sold quantity, so they can compute excess supply, but not excess demand. For this reason, the price adjustment process takes place step by step, not instantaneously, and it is based on adaptive expectations.
4. Learning process. In this model, at each time t, each firm collects the available information about the current state of a subset of other agents, those who lay on the same production stage, and starts its own decision-making process, to set price and quantity for the next period $(t + 1)$. In this way, the dynamics of this kind of models is backward-looking, sequential and path-dependent.
5. Non-linearity. The interactions that occur in this model are inherently non-linear. Firms choose to trade with different partners at each step. It is also possible that, at a certain time, some firms do not sell nor buy anything. Firms on each level choose to buy input from the firm on previous level with the minimum price. Thus, it may happen that firms with higher prices do not have demand for their

output, so they do not sell. This means also that they will not buy input to produce additional output in the next period, because of their inventories full of unsold products.

Notation, Assumptions and Initial Conditions

Firms are organized in a vertical supply chain with S production stages and n_s firms active at each stage. Each firm in the network is indexed by $(s;i)$,where $s = 1, \ldots, S$ is the production stage, and $i = 1, \ldots, ns$ is the position of the firm in that stage. Figure 5.1 shows an example of the supply chain.

Final consumers are represented by a linear inverse demand function:

$$\tilde{P}\left(Q^d\right) = a - bQ^d$$

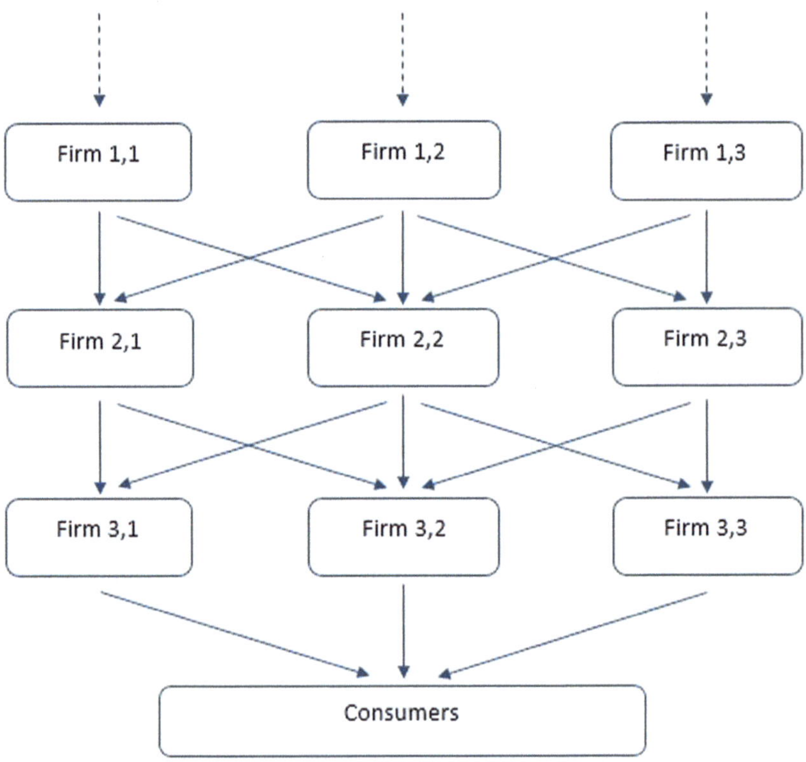

Fig. 5.1 Supply chain example

where $\tilde{P}$ is the minimum price among last stage firms' prices, $\tilde{P} = \min\left\{p_{S,1}, p_{S,2}, \ldots, p_{S,n_S}\right\}$, and a and b are parameters exogenously defined. Thus, we simply get aggregate demand for retail level:

$$Q^d = \frac{a - \tilde{P}}{b}$$

Each firm has the simplest production function, using only one homogeneous input and transforming it into a homogeneous output with a one-to-one technology: $y_{s,i} = x_{s,i}$, where y is the output produced (and offered), and x is the quantity of the unique input purchased.

Each firm has an inventory, $I_{s,i}$, where output produced is stored. After the selling process, quantity eventually unsold is stored too. There are no storage costs, and the good is supposed to be not perishable.

Each firm has a simple cost structure, as the following: $C_{s,i} = k_{s,i} \cdot x_{s,i}$, where $k_{s,i}$ is the input price set by the supplier. If the firm bought the amount of input needed from two different suppliers (say 1 and 2), with two different prices, obviously, cost function will be: $C_{s,i} = k^1_{s,i} \cdot x^1_{s,i} + k^2_{s,i} \cdot x^2_{s,i}$. We assume there are no fix costs.

After trading, firms compute their profits: $\pi_{s,i} = p_{s,i} \cdot q_{s,i} - k_{s,i} \cdot x_{s,i}$, where $q_{s,i}$ is the quantity of output sold. If $y_{s,i} > q_{s,i}$, the difference between the quantity produced and the quantity sold is stored in the inventory, so $I_{s,i} = y_{s,i} - q_{s,i}$ and, at each time t, $I_{s,i;t} = I_{i,s;t-1} + y_{s,i} - q_{s,i}$.

At the beginning of the experiment, we make the following assumptions:

Assumption 1 Firms on the upstream stage $s = 1$, have different marginal costs $k_{1,s}$, exogenously assigned. $k_{1,i}$ is a random variable normally distributed with mean μ_k and variance σ_k: $k_{1,i} \sim IIN(\mu_k, \sigma_k)$.

Even if the input is homogeneous, we assume that asymmetric information and search costs imply that, in this initial phase, firms may end up with different marginal costs.

Assumption 2 At time $t = 0$, before trading takes place, there is local interaction among firms. This means that each firm has a unique supplier, randomly assigned among firms on the previous level. Having different suppliers, so different marginal costs, firms will set different prices.

Again, we assume that the very first-time firms interact, there is incomplete information, so firms on downstream levels, $s > 1$, may select a supplier even if its price is not the lowest. Thus, initial conditions consist in Bertrand equilibrium with asymmetric costs. Firms at each stage set the following price, $p_{s,i} = k_{s,i}$, and offer the following quantity:

$$y_{s,i} = \frac{a - k_{s,i}}{b(n_s + 1)}.$$

By Bertrand surely the second minimum would dominate, but due to lack of capacity, other higher prices may survive. The supply capacity rationing implies a certain degree of market power for residual firms filling on their own local market. Local interaction occurs only at $t = 0$, when firms set their price and quantities for the very first time. When "real trading" starts, each firm chooses its supplier among all firms on the previous stage.

Trading Process

Given initial conditions for $p_{s,i}$, and $y_{s,i}$, the evolution of the system is observed in discrete time steps, $t = 1, 2, \ldots T$. We assume, at each time t, trading process to occur backward, so it starts with the transactions between retailers and final consumers and goes on in a backward sequential manner, stage by stage, from downstream to upstream. Starting from the retail level, where $s = S$, final consumers, looking at the retailers' prices, set their demand Q^d. In trading there is no bargaining: consumers can observe prices of each .rm at this level, and, since the good is homogeneous, they want to purchase the final good from the firm who shows the minimum price, according to the so-called minimum price rule. However, this minimum price .rm is capacity constrained, so there will be rationing. If the quantity demanded from consumers is greater than the quantity offered by the firm with the minimum price, that firm will sell all its output, and consumers will demand the residual to the firm with the second minimum price, and so on, until quantity demanded will become zero. Whenever a firm sells some quantity, it realizes revenues, simply calculated as the product between quantity sold and price. If a firm doesn't sell all the quantity it supplies, the difference between sold and supplied quantity, excess supply, become inventories.

After having sold, the i-th firm at retail level adjusts its control variables, i.e. the output price and the potential supply for the next period. This is the process of "learning", in which firms collect the available information about the current and past state of an agents' subset and, based on this information and their own state, they "update" their strategic variables, according to their behavioural rules, in order to adapt to changing business conditions. This process will be explained in the following section. When firms complete their learning process, they know the price they will set, the quantity they will offer, thus their demand for input.

When firms at retail level formulate their demand for input, trading between these firms and those ones at the previous stage takes place. Again, we do not allow for bargaining among firms. Buyers can observe all their suppliers but want to purchase from the one who shows the minimum price; when she ends capacity, buyers will ask to the second last firm, until all firms demand is satisfied. Note that buyers act one at a time, picked randomly: in this way the firm that succeeds in buying from the lowest price supplier is always different. Once they have purchased input, they transform it into their final good (according to their production function).

Buyers incur in costs equal to the purchased quantity times the price they paid. (If they buy different quantity from different suppliers, the total cost is the sum between the different purchased quantities, each multiplied by its paid price). Sellers realize revenues and start their learning process, to make new decisions about price and quantities to refill the store.

The process of trading goes on in a backward sequential manner: the i-th firm who lays on the production stage s first sells on to firms who lay on the stage $s + 1$ and, after having "learned" the quantity she wants to produce, she purchases from firms on stage $s - 1$. This process stops when the firms at the first stage upstream are sellers. They do not become "buyers", meaning they do not exchange with other agents. They update their storehouses, purchasing from the "outside world" the quantity of input they desire, and incur in costs equal to k_i exogenously assigned. When every firm in every production stage has updated its inventory and fixed its new price, this new set of microeconomic variables is fed into the economy for the next-step iteration. At the end of each time t, every firm is characterized by price, quantity offered, quantity sold, quantity of input purchased, variable costs, revenues and profit.

Learning Process

In complex interactive systems, individual decision processes are unavoidably adaptive. Expectations on the future and results are clearly an important part of the decision-making process, but foresights are taken over finite horizons and are modified sequentially in the light of realized outcomes. Strategic variables, i.e. price and quantity, are updated according to "error-correction" procedures. In our model, at every t, after having sold, each firm must decide its $p_{s,i}(t + 1)$ and $y_{s,i}(t + 1)$ she wants to offer the next period. Agents take their decisions after have met their demand: this means that here we are imaging a "demand-driven" market, where production is directly triggered by the demand itself.

In designing the learning process, we follow Delli Gatti et al. (2011), and we assume that a firm can change either the price or the quantity, but not both at the same time. (In other words, the strategies consisting in "changing the price" and in "changing the quantity" are mutually incompatible). Therefore, if the i-th firm at time t decides to set a new price $p_{t+1} \neq pt$, then the supply will be the same, $y_{t+1} = y_t$, and vice versa; if she decides to offer a different quantity $y_{t+1} \neq yt$, then the price will not change $p_{t+1} = p_t$.

Firm decisions depend on signals coming from the internal condition and from other competitors. The information set, relevant for price or quantity adjustment of the i-th firm at each time t, consists of two components:

- $I_{s,i} = y_{s,i} - q_{s,i}$: their own inventory. At the end of the trading process, if the firm ends up with a positive inventory, she argues there is a situation of excess supply. Note that this excess can also be due to unsold quantity in previous periods;

however, firms aim to sell all its storage products. If, otherwise, the inventory is empty, firms are not able to recognize if demand happened to be equal to supply or if there was excess demand. Firms can observe only quantity sold, but not the amount of demand, so they are not able to recognize and eventually quantify excess demand.

- $P_{s,i} - \underline{P}_s$: the deviation of the individual firm's price from average price of firms who lay on the same level, i.e. firm's direct competitors. Even if the "relevant price" in this model is the lowest price, firms are not able to observe individual prices which are private knowledge. They observe the average price and use it to form their expectations. If the deviation from its price and the average one is positive, the firm recognizes that she is charging a price higher than its competitors, so she may be "out of business", and vice versa; if this distance is negative, the firm is charging a price lower than its competitors.

Internal conditions (i.e. the level of the inventory and the individual price) are private knowledge, while the average price is common knowledge. The combination of these two conditions produces four different cases, which correspond to four different behavioural strategies, but all of them have the same objective: maximizing the profit function.

1. If $I_i = 0$ and $p_{s,i}\ ''\ \underline{P}_s$, (where $\underline{P}_s$ is the average price of firms in the production stage s). The empty inventory signals an excess demand, even if not quantifiable: in the next period, the firm will choose to offer the same quantity sold this time, but at an increased price, in order to gain more profits.
2. If $I_i > 0$ and $p_{s,i} > \underline{P}_s$, the firm recognizes that she is charging a too high price, which causes supply excess. In this case, the firm is induced to reduce the price, trying to attract a larger portion of customers, leaving the quantity unchanged.
3. If $I_i = 0$ and $p_{s,i} > \underline{P}_s$, the firm probably is facing a demand excess even if her price is higher than her competitors. Therefore, she expects that demand will increase next period, so she chooses to produce a higher quantity, which she will try to sell at the same price.
4. If $I_i > 0$ and $p_{s,i}\ ''\ \underline{P}_s$, the firm recognizes that there is a supply excess, even if her price is lower with respect to the average. Therefore, she decides to produce a lower quantity, leaving the price unchanged.

Formally, we can define the following rules. We drop the notation of the production stage s, since these rules are valid for each stage. Pricing rule is:

$$p_{i,t+1} = \{\left\{p^C_{i,t+1}; p_{i,t}\left(1+\varepsilon_{i,t}\right)\right\} if\ I_i = 0 \text{ and } p_i \leq \underline{P} \left\{p^C_{i,t+1}; p_{i,t}\left(1-\varepsilon_{i,t}\right)\right\} if\ I_i > 0 \text{ and } p_i > \underline{P} \quad (5.1)$$

where $\varepsilon_{i,t}$ is an idiosyncratic random variable uniformly distributed on the support $(0; h_\varepsilon)$. $p^C_{i,t}$ is the minimum level of price the firm has to charge to cover average costs. Therefore, there exists a lower bound to the reduction of price, represented by the budget constraint:

$$p_{i,t}^{C} = \frac{k_{i,t} \cdot x_{i,i}}{q_{i,t}}$$

Output rule is the following:

$$y_{i,t+1} = \{ q_{i,t}\left(1+\eta_{i,t}\right) if\ I_i = 0 \,\text{and}\, p_i > \underline{P}\, q_{i,t}\left(1-\eta_{i,t}\right) if\ I_i > 0 \,\text{and}\, p_i < \underline{P} \quad (5.2)$$

where $\eta_{i,\,t}$ is an idiosyncratic shock variable uniformly distributed on the support $(0; h_\eta)$. Here, $y_{i;\,t+1}$ represents the quantity of input the i-th firm will demand to its supplier. Note that quantity adjustments are based on quantity sold. If a firm hasn't sold anything, she will not buy anything, because she has the inventory full of output unsold. So, quantity actually offered at each time is quantity produced in that time (which depend on quantity sold last time) plus the inventory. There is an upper bound to the quantity: firms cannot produce a quantity that is greater than their own capacity.

Equations 5.1 and 5.2 represent a kind of firms' reaction function, so these are their best responses to competitors' behaviour given internal conditions. Through these strategies, firms aim not to have inventories with unsold quantity nor to set prices out of business, so they aim to maximize their profit function. Adjustment strategies available for i-th firm at time t can be summarized by Fig. 5.2 (Delli Gatti et al. 2011; Bhaskar et al. 1993).

In the region characterized by $p_{s,i}$ " $\underline{P}_s$ and $y_{s,\,i} < q_{s,\,i}$, firm has an incentive to increase the price: all produced output has been sold, so the agent wants to increase its revenue. In principle she has an incentive to increase quantity produced too, but by increasing the individual price, the firm will lower demand in the future, and the absorption of the increased volume of output is not granted. Otherwise, if the firm has sold all produced quantity, even though she had a price greater than its competitors, she has an incentive to increase supply.

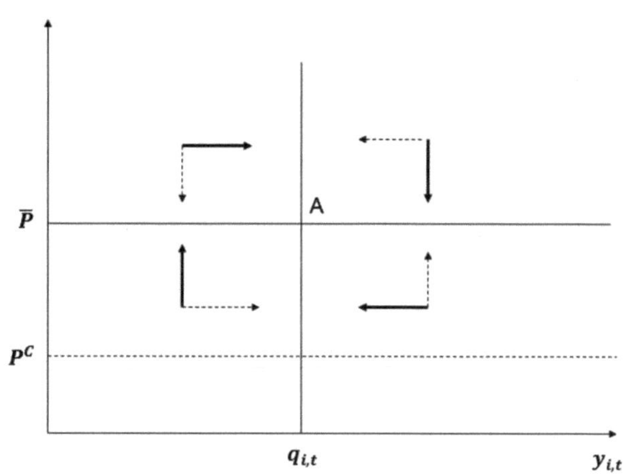

Fig. 5.2 Price and quantity adjustments for a generic firm i

In the same way, firm adjusts only one strategic variable when it lays on the other two regions of the figure. This graphic representation shows that there is an implicit tendency for the firms to move towards an equilibrium and that this tendency is characterized by a spiralling pattern on the price-quantity space. However, it is necessary to consider that the equilibrium itself changes over time, evolving according to the evolution of firms' behaviour.

Note that in this learning process, firms do not take into account their own past stories. Furthermore, we do not consider fixed costs and we assume there isn't technological innovation. This can be justified because we are implementing a short-run equilibrium analysis, so we will find a short-run equilibrium.

Results

Using the software NetLogo (version 6.0), we run several sets of simulations with different hypotheses on the market structure. Experiments differ by the organization of the production chain: we assume different numbers of firms per stage and different firms' behaviours. Here, we present simulations' outputs.

We first run a baseline scenario, one for each production chain setting, characterized by a different number of active firms at each stage. By doing so, we analyse price and quantity dynamics, in order to appreciate the effects of the minimum price rule, when firms are capacity constrained in supply. In a second phase, we assume the presence of an agreement (cartel) between a given number of firms at a certain production stage. Following the standard collusive behaviour, cartel members fix a price above the competition level, but, according to our assumptions, cartel members remain price-takers in the market for their inputs, and there are no specific efficiency gains. We analyse cartel's effects on price dynamics and on profits, considering different cartels with different numbers of members. Throughout all experiments, the set of parameters listed in Table 5.1 have fixed values.

Table 5.1 Parameters values

	Parameter	Value
T	Number of time periods	200
a	Demand function parameter	100
b	Demand function parameter	1
μ_k	Mean of $k_{1,i}$ (input cost for firms at the upstream production stage)	15
σ_k	Variance of $k_{1,i}$	5
h_ε	Maximum growth rate of prices	0,03
h_η	Maximum growth rate of quantities	0,03

The Baseline Scenario

We first construct a very simple scenario, considering a supply chain with $S = 4$ production stages, and $n_s = 3$ firms active at each stage. In this simple setting, we are interested in understanding price, quantity and profit dynamics, in the so-called baseline scenario, where firms compete according to the minimum price rule, given capacity constraints.

Throughout the simulation, we assume that, starting from consumers, demand will try to buy everything at the minimum price. Capacity rationing pushes demand to the second low price and so on. Then sellers have to buy inputs following the same mechanism, etc. Due to this minimum price rule, price obviously is the strategic variable, but since firms are capacity constrained, quantity becomes a strategic variable too. Selling firms may experience excess supply but are not able to infer excess demand. Because of this asymmetry, when they have to update strategic variable, they rely on the available information, i.e. deviation of their own price from the average price of firms who lay on the same level, and on quantity eventually unsold and stored in the inventory. Thus, each firm forms its adaptive expectations and takes its decisions according to them. All firms aim to maximize profit, and this goal implies not to have quantity unsold in inventories and set a profitable price to attract customers. Therefore, they update selling price and supply following the rules exposed in the previous section. Note that in this model there is no bargaining between parties: the price is unilaterally set by suppliers.

We show what happens "on average", i.e. averaging prices, quantity and profits, on the number of existing firms per layers, studying the evolution of these average variables over the simulation steps. Having four layers, we have an average path for

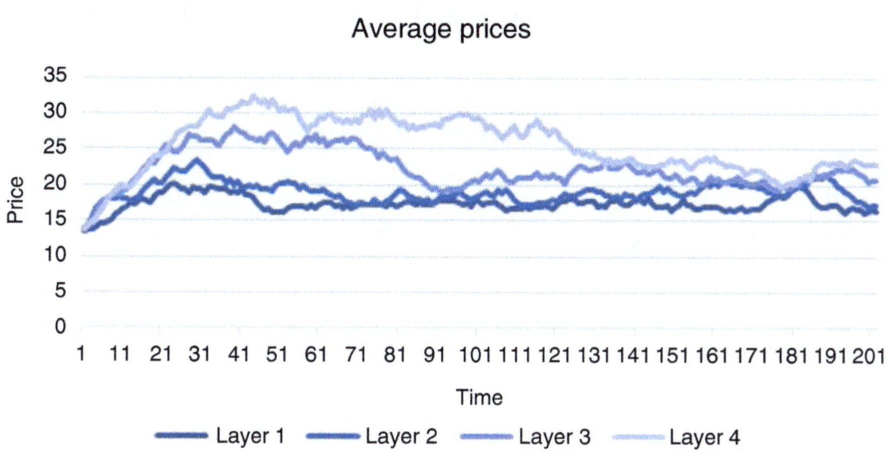

Fig. 5.3 Average prices on each layer, when $n_s = 3$

each layer. In Fig. 5.3, we observe that average prices initially increase with respect to initial cost values, k. The supply capacity rationing implies a certain degree of market power for all firms, who can set a mark-up above costs. This means that prices slightly increase, above the perfect competition level (i.e. marginal cost). Then, competition on the minimum price seems to keep prices stable.

Figure 5.4 shows a constant and linear trend of growth for quantities. However, both for prices and quantities, we do not find a fixed value of equilibrium, as if we computed it analytically: continuous interactions among agents determine an equilibrium with fluctuations around a certain trend. Maximizing behaviour shows that moving price and quantity causes average profits to monotonically increase (Fig. 5.5).

Comparing these results with a context characterized by a larger number of active firms, $n_s = 6$, (leaving unchanged the number of production stages $S = 4$), we can observe the stronger competitive pressure effects, especially on prices. In fact, we show the same average variables when there are $n_s = 6$ firms. From Fig. 5.6, it is possible to observe deeper prices fluctuations, whilst Fig. 5.7 shows lower profits than the previous scenario. Note that firms on layer 1 upstream reach higher profits because we don't assume there exist capacity rationing in raw materials. However, they are higher when there is a greater number of active firms, so when the competition pressure grows.

The Cartel Scenario

Here we assume the presence of a cartel, i.e. an illicit agreement between firms, whose aim is to increase their collective profits maintaining prices at a high level, and consequently restricting competition. As stated in the Art. 101 TFEU, illicit

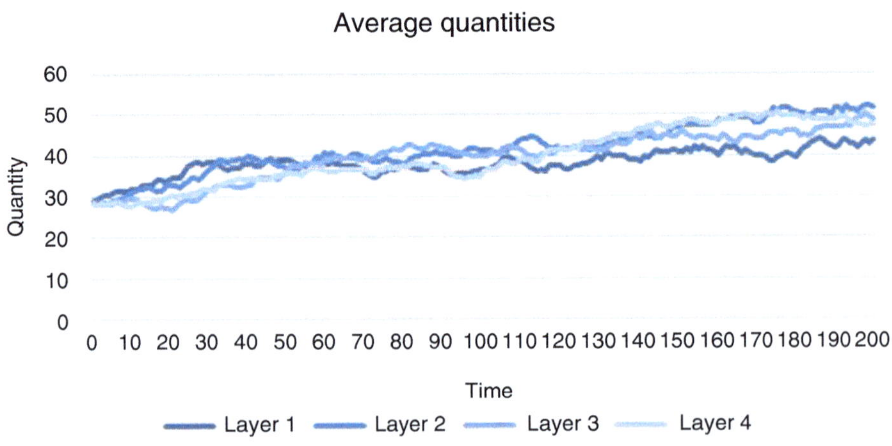

Fig. 5.4 Average quantities on each layer, when $n_s = 3$

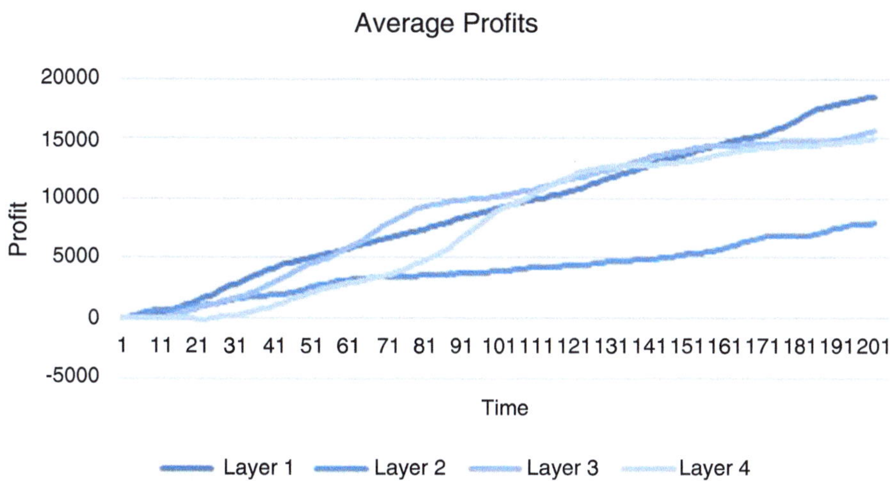

Fig. 5.5 Average profits on each layer, when $n_s = 3$

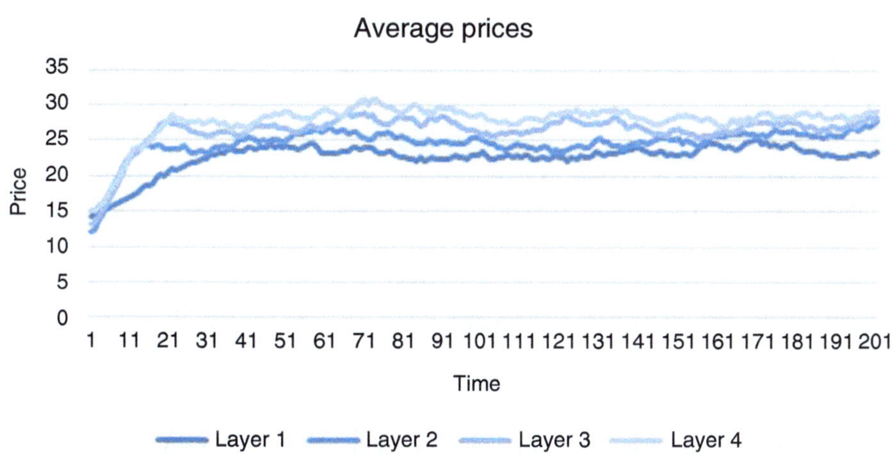

Fig. 5.6 Average prices on each layer, when $n_s = 6$

agreements (but also decisions and concerted practices) include not only those which fix higher prices but also those which limit or control production, markets or investment and implement other conducts which can prevent or restrict the competition. However, our setup implies that cartel members only increase their prices: they remain price-takers in the market for their inputs, and there are no specific efficiency gains that would somehow allow the cartel to produce at lower costs than its members could in competition.

In this context, we suppose that at a certain time during simulation, on layer $s = 2$, a number m of firms, with $m \in 2\ [2; n_s]$, increase their prices, setting them at the standard collusive level due to linear demand and cost functions, while competi-

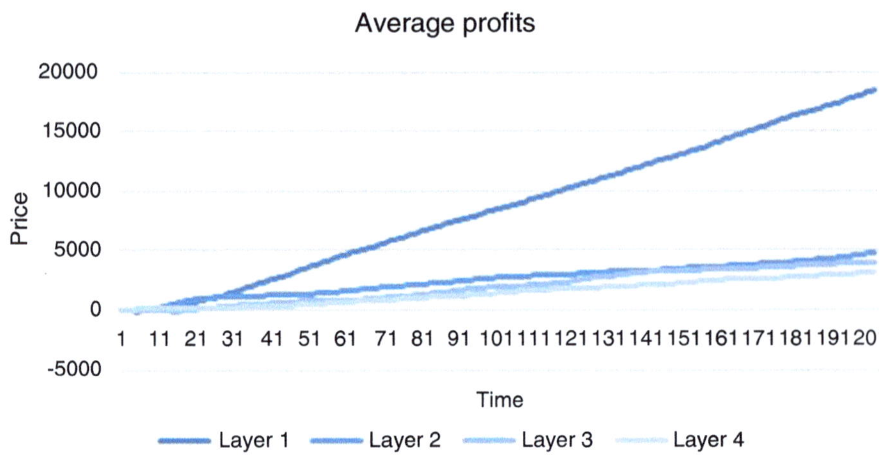

Fig. 5.7 Average profits on each layer, when $n_s = 6$

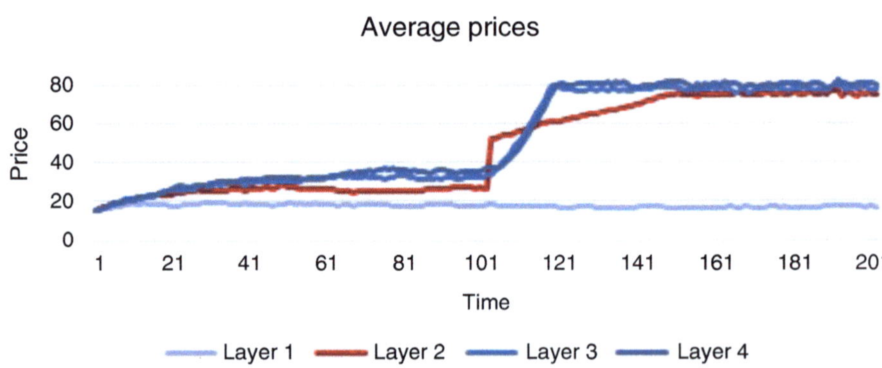

Fig. 5.8 Average prices on each layer, when $m = 3$

tive conditions in all other layers remain the same. As the following figures show, the cartel price will be significantly higher than the one resulting from the competitive mechanism supposed until now.

All cartel members, during the entire cartel period, set the following price:

$$p^C = \frac{a + \sum_{i=1}^{m} k_i}{2}$$

Cartel members, once their agreement takes place, keep on fixing the collusive price, so they do not perform the classic learning process of this model. This means that, even if they do not observe higher profits as they expected, they do not adjust their price according to their own inventory and their competitors' price, given that they must comply with the agreement they made. Regarding quantity choice, we

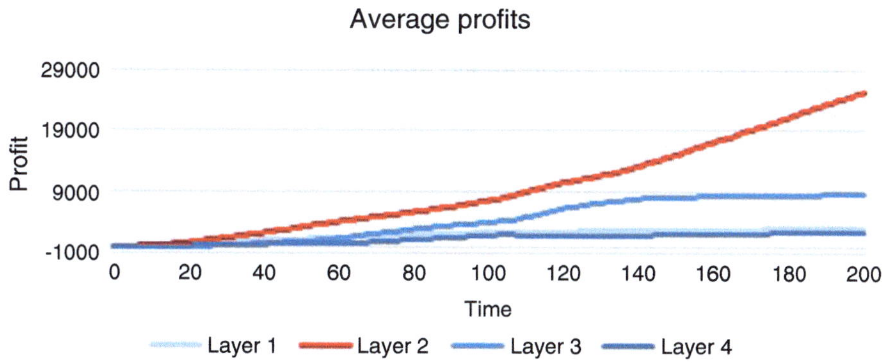

Fig. 5.9 Average profits on each layer, when $m = 3$

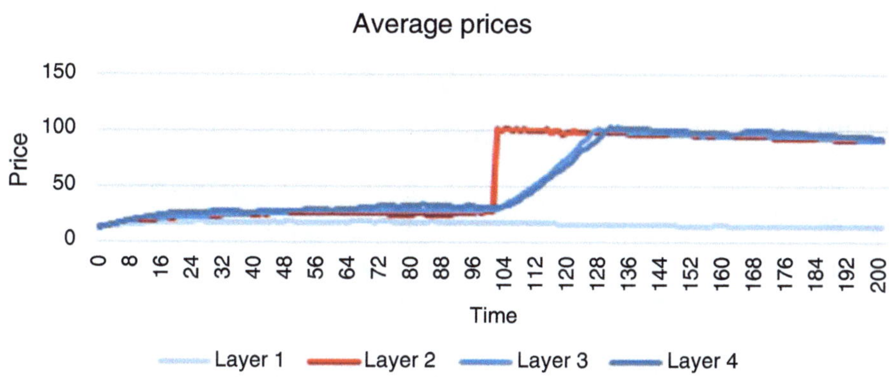

Fig. 5.10 Average prices on each layer, when $m = 6$

assume that cartel members slightly increase the quantity they produce, in such a way as to expand and serve a greater part of the demand they face. We will see that this behaviour will not determine the expected higher profits.

We consider a supply chain with $S = 4$ layers, and $n_s = 6$ active firms on each layer, assuming that cartel occurs on layer $s = 2$; that it starts after simulation for the baseline scenario stabilizes, i.e. at time $t = 101$; and that it will be defined on the second half of the simulation. Figures 5.8 and 5.9 show, respectively, average prices and average profits when three firms on layer 2 collude. Red line represents layer 2, where the cartel occurs. As can be seen, the average price of layer 2 drastically grows when the cartel starts, causing the growth of the downstream layers' prices too. Downstream firms, facing a higher marginal cost, rise in turn their final price: this means that price increase due to the cartel surely has been transmitted downstream to the chain. Profits of cartel layer reach a higher level than the other ones.

When all firms on layer 2 collude, results, as shown in Figs. 5.10 and 5.11, are completely different from the previous situation. It can be noticed that firms on layer 3, who are the direct purchasers of the cartel, earn negative profits during all

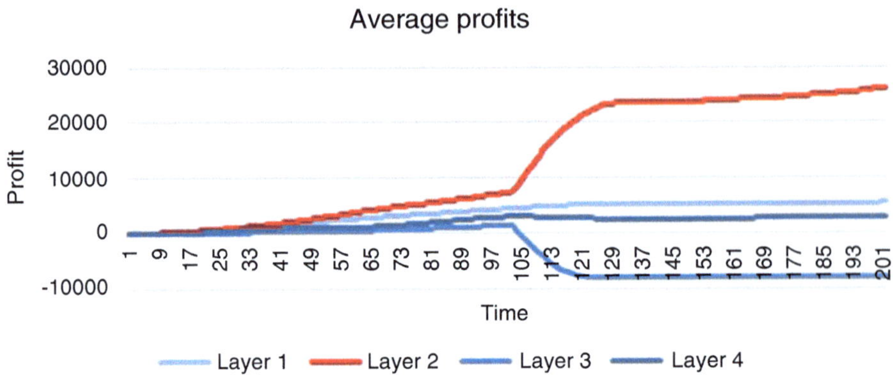

Fig. 5.11 Average profits on each layer, when $m = 6$

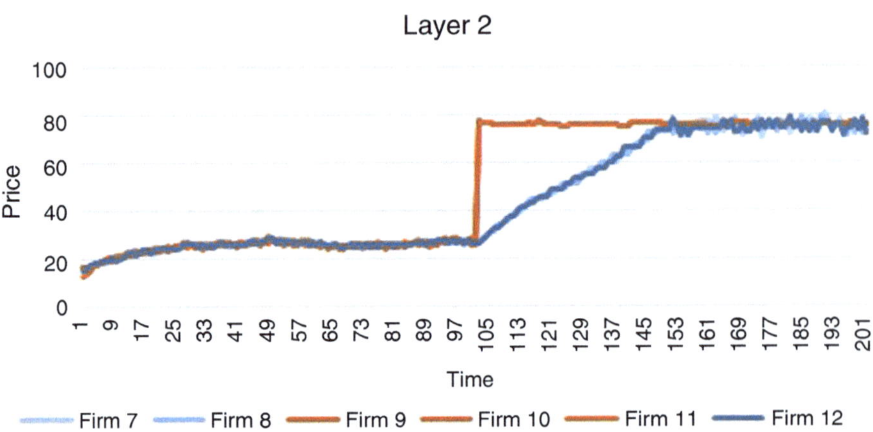

Fig. 5.12 Prices of individual firms on layer 2, when $m = 3$

the second half of the simulation because they cannot choose a supplier different from the cartel. In this situation, they are not able to cover their higher marginal cost, despite the budget constraint that operates during the learning process. This probably means that cartel members are fixing a too much high price that probably in a real contest would not meet the demand. Indirect purchasers of the cartel, instead, can reach positive values of profits.

Interesting results come from individual firm variables analysis when only three firms collude. Figure 5.12 shows how cartel members' behaviour (red lines) causes an incentive to increase price also for outsiders (blue lines), i.e. firms who do not collude. The higher cartel price makes the other firms' prices raise too. Furthermore, Fig. 5.13 shows profits dynamics that could seem counterintuitive: firms outside the illicit agreement gain higher profits, with respect to firms who collude. This is a consequence of the cartel members' behaviour: they keep on fixing the same price

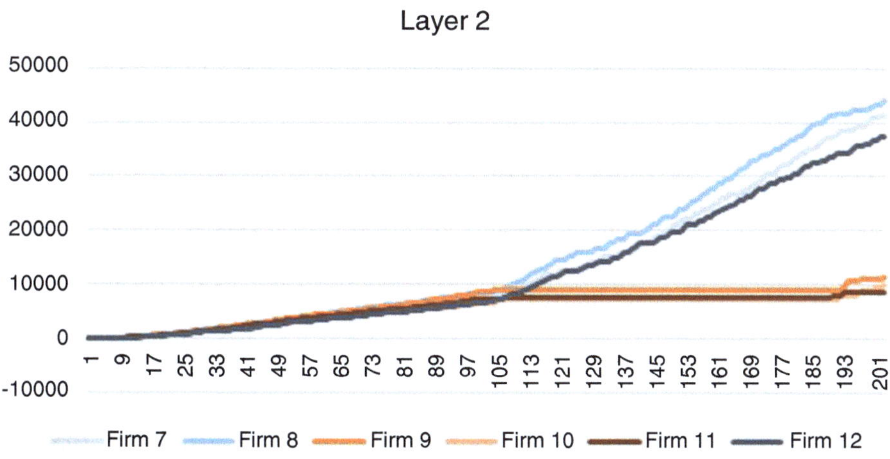

Fig. 5.13 Profits of individual firms on layer 2, when $m = 3$

established from the agreement. This means that they stop controlling for quantity sold. Outsiders, on the contrary, see what is happening and take advantage of this situation. They understand that there is the possibility to raise price, but, at the same time, they pay attention to quantity sold: by doing so, they succeed in earning higher profits.

Price Increase and Pass-on

Here we focus on prices, to analyse in detail price increase, due to the illicit agreement, and how it is transmitted downstream to the chain. We first compute the average price increase for each layer, as the percentage difference between the average price of all firms considering only the period before the cartel ($t = 1, \ldots 100$) and the average price of all firms considering only the periods after the cartel took place ($t = 101, \ldots 200$). In Table 5.2, we report these average price increases computed for different numbers of cartel members (m). When the number of colluding firms grows, price increase grows too, and this is true both for layer 2, where the cartel takes place, and for layers downstream the chain (layer 3 and 4).

Then we compute pass-on rate, as the ratio between final price increase and marginal cost increase (i.e. input price). Results, reported in Table 5.3, show that firms on layer 3 pass a smaller proportion of the cost increase they face the greater the number of cartel members. This result is in line with what the theory predicts: pass-on rate increases when competition level is higher, which means, in this case, when a smaller number of firms collude. It is also possible to observe that as the "distance" from the cartel layer grows, the pass-on rate decreases, so firms downstream pass a smaller proportion of the cost increase they face.

Table 5.2 Price increase for different numbers of cartel members

m	$\underline{P}_2$ (average price layer 2)	$\underline{P}_3$ (average price layer 3)	$\underline{P}_4$ (average price layer 4)
2	1455	1442	1283
3	1864	1720	1533
4	2306	1934	1712
5	2840	2223	1937
6	3478	2524	2225

Table 5.3 Pass-on rates

m	$\underline{P}_3 / \underline{P}_2$	$\underline{P}_4 / \underline{P}_3$
2	0,991	0,889
3	0,923	0,891
4	0,839	0,885
5	0,783	0,871
6	0,726	0,882

Conclusions

In the context of cartel damages action, national courts of EU Member States have the complex task to quantify actual damages due to an antitrust infringement. This task includes estimating the pass-on rate. Here, we developed a model in order to study a vertical supply chain, producing a homogeneous good with homogeneous inputs at each stage of production. Thanks to the ACE approach, we can appreciate how direct endogenous interactions among heterogeneous agents determine an equilibrium path. We consider that both price and quantity are the strategical variables, and competition acts according to the "minimum price" rule, with firms that are capacity constrained. Results show that under linear demand and cost functions we find a fluctuating equilibrium path. Then we allow for the possibility to collude: during simulation, a variable number of firms form a cartel and decide to fix a higher price, above the competition level. This illicit agreement changes the market structure, causing a different equilibrium path.

Interesting results are the following:

1. Cartel price represents an incentive for outsiders to raise their prices too.
2. Price increase applied by the firms of the cartel and the firms out of the cartel grows as the number of cartel members increases, and this is true also for layer downstream the chain.
3. Pass-on rate decreases as the market concentration increases.

This model, however, considers only cartel downstream effects, but it should be analysed also what happens upstream if the cartel could use some buyer power. Further investigations could be aimed to show how equilibrium reacts to demand shock. It can be interesting also to model other topologies of the network, introducing geographical differentiation, according to which a firm on a certain level cannot buy from all firms of the previous level but only from someone of them (with whom it has a link); we introduce geographical differentiation. It could be possible to introduce different competitive settings at different production stages. For example, we can consider a situation with a great number of retail firms, so a very competitive market, but a monopoly at an intermediate level, is a bottleneck. Here, we can see what happens when at an intermediate level of the production chain there is a monopsony power (buyer power), and how this will change prices. An example of this scenario is food production where downstream there are supermarket distribution chains.

References

Ashenfelter, O., Ashmore, D., Baker, J., & McKernan, S-M. (1998). *Identifying the firm-specific cost pass-through rate*, Federal Trade Commission Working Paper No. 217.

Ashton, D., & Henry, D. (2013). Competition damages actions in the EU – Law and practice, Elgar competition law and practice series.

Bhaskar, V., Machin, S., & Reid, G. C. (1993). Price and quantity adjustment over the business cycle: Evidence from survey data. *Oxford Economic Papers, 45*(2), 257–268.

Bonnet, C., & Villas Boas, S. (2013). *An analysis of asymmetric consumer price responses and asymmetric cost pass-through in the French co¤ee market*, Working Paper No. 10, Transparency of Food Pricing project.

Davis, P., & Garcés, E. (2009). *Quantitative techniques for competition and antitrust analysis*. Princeton University Press. Princeton, New Jersey

Delli Gatti, D., Desiderio, S., Gaffeo, E., Cirillo, P., & Gallegati, M. (2011). *Macroeconomics from the bottom up*. Milan: Springer.

European Commission. (2013). Practical guide on quantifying harm in actions for damages based on the breaches of article 101 or 102 of the treaty on the functioning of the European Union, 11 June 2013, SWD (2013) 205.

European Commission. (2016). Study on the passing-on of overcharges, Final report, http://ec.europa.eu/competition/publications/KD0216916ENN.pdf

Friederiszick, H. W., & Röller, L. H. (2010). Quanti…cation of harm in damages actions for antitrust infringements: Insights from German cartel cases. *Journal of Competition Law and Economics, 6*(3), 595–618.

Han, M. A., Schinkel, M. P., & Tuinstra, J. (2009). *The overcharge as a measure for antitrust damages*, Amsterdam Center for Law & Economics working paper, 2008–08.

Hellwig, M. F. (2006). Private damage claims and the passing-On defense in horizontal price-fixing cases: An economist's perspective, Max Planck Institute for Research on Collective Goods Preprint, 2006/22.

Kim, D., & Cotterill, R. W. (2008). Cost pass-through in differentiated product markets: The case of U.S. processed cheese. *Journal of Industrial Economics, 55*(1), 32–48.

Leibtag, E., Nakamura, A. O., Nakamura, E., & Zerom, D. (2007). Cost pass-through in the U.S. Co¤ee Industry, economic research report No. (ERR-38) 28.

Miller, N. H., Osborne, M., & Sheu, G. (2016). Pass-through in a concentrated industry: Empirical evidence and regulatory implications, Manuscript.

RBB Economics. (2014). *Cost Pass-Through: Theory, measurement and policy implications*, Report for the O¢ ce of Fair Trading.

Rochet, J., & Tirole, J. (2011). Must-take cards: Merchant discounts and avoided costs. *Journal of the European Economic Association, 9*(3), 462–495, 06.

Tesfatsion, L. (2005). Agent-based computational economics: A constructive approach to economic theory. In *Handbook of computational economics*. Elsevier, Amsterdam, North Holland. (Vol. 2, pp. 831–880).
Tesfatsion, L., & Judd, K. (2006). Agent-based computational economics. In *Volume 2 of Handbook of computational economics*. Amsterdam: North Holland.
Tirole, J. (2003). *The theory of industrial organization*. Cambridge, MA: The MIT Press.
Verboven, F., & van Dijk, T. (2009). Cartel damages claims and the passing-on defense. *Journal of Industrial Economics, 57*(3), 457–491.
Weyl, E. G., & Fabinger, M. (2009). Pass-through as an economic tool, Manuscript.

Chapter 6
Modeling the Dynamics of Reward-Based Crowdfunding Systems: An Agent-Based Model of Kickstarter

Marco Campennì and Federico Cecconi

Abstract In this chapter, we present an agent-based model of Kickstarter, maybe the most successful crowdfunding platform available nowadays. We aim to use the model to better understand cognitive strategies and behavioral dynamics governing Kickstarter, considered as a multi-actor system where agents adopting different strategies interact. The model shows how the success of agents is affected by the strategy they adopt and provide interesting insight about non-linear dynamics resulting from the interactions among agents.

Keywords Kickstarter · Agent-based modelling · Social network analysis · Profiling investors · Cognitive strategies

Introduction

Among the many (slightly) different definitions of crowdfunding proposed over the last decade, in this work we adopted the definition suggested by Belleflamme et al. (2010), stating that crowdfunding involves an open call, essentially through the Internet, for the provision of financial resources either in form of donation or in exchange for some form of reward and/or voting rights.

Crowdfunding is a relatively new method for funding a variety of new ventures. It allows individual campaign promoters of for-profit, cultural, or social projects to request funding (i.e., money) from many other individuals (i.e., backers), often in return for future products or equity.

The range of crowdfunding projects may considerably vary in both the type of goal(s) and the amount established as the "target goal" of the specific campaign.

M. Campennì (✉)
Biosciences, University of Exeter, Penryn, Cornwall, UK
e-mail: m.campenni@exeter.ac.uk

F. Cecconi
LABSS ISTC-CNR, ROME, Italy

© Springer Nature Switzerland AG 2019

F. Cecconi, M. Campennì (eds.), *Information and Communication Technologies (ICT) in Economic Modeling*, Computational Social Sciences,
https://doi.org/10.1007/978-3-030-22605-3_6

The goal of the campaign may be promoting small artistic projects, or the campaign may represent an alternative way for (traditional) entrepreneurs to seek hundreds of thousands of dollars in seed capital (Schwienbacher and Larralde 2010).

Over last two decades, we assisted to an impressive increase in the interest in crowdfunding.

Some billions dollars spent by millions of individual crowdfunding backers, and large-scale actions by the US Congress to encourage crowdfunding as a source of capital for new ventures (Martin 2012).

Despite the massive societal and economic impact of the phenomenon on the way we design, plan, and implement the production of goods, there is still a considerable lack in the knowledge and understanding of the dynamics of crowdfunding, and the scientific research and literature on it seem to be limited.

Scholars know very little about the dynamics of successful crowdfunding and cognitive mechanisms and strategies governing campaign promoters and backers behaviors, as well as the general distribution and use of crowdfunding mechanisms.

It is still unclear whether the crowdfunding phenomenon will confirm or contradict existing theories about how ventures raise capital and achieve success. It is definitely also unclear what the long-term implications of crowdfunding will be.

In this chapter, we present an agent-based model of Kickstarter (see Kickstarter Wikipedia webpage (2015)), maybe the most successful crowdfunding platform available nowadays.We aim to use the model to better understand cognitive strategies and behavioral dynamics governing Kickstarter, considered as a multi-actor system where a specific typology of agent can promote a campaign (i.e., campaign promoter agents), another typology of agent can advertise such a campaign (i.e., Kickstarter), and finally, another typology of agent can invest in such a campaign (i.e., backer agents).

The Model

In this work, the model we developed aims at clarifying the dynamics of crowdfunding considering a specific case study, i.e., Kickstarter, where three types of agents involved in transactions, namely, backers, campaign promoters, and the platform used, to (i) promote campaigns and (ii) collect funds interact (see Bouncken et al. 2015 for an interesting analysis of crowdfunding as three types of agents interaction).

We decided to model the Kickstarter behavior because of its relevance in the landscape of different online crowdfunding platforms and because of the considerable amount of funds it is able to attract and manage. Our main interest is in better understanding the specific (behavioral) dynamics resulting from the interaction of all parties involved and the cognitive strategies different type of agents adopt while interacting with other agents.

In the model three different types of agents are implemented: backers, campaign promoters, and Kickstarter platform. Using this model, we can investigate the complex, potentially non-linear dynamics emerging from interactions between campaign promoters (i.e., agents looking for resources to fund their projects), backers (i.e., agents potentially funding other agents' campaigns), and Kickstarter (i.e., the environment and the environmental conditions where behaviors and dynamics occur).

Kickstarter as we modeled can be interpreted as a social institution in game theory terms: it is the combination of (i) the set of rules of a (strategic) game played by (three) different types of agents and (ii) the resulting equilibria (Hindriks and Guala 2015). Because of the complex interplay resulting from the interactions between agents adopting different strategies, we were unable to provide an analytical definition and description of the model, and therefore we designed, developed, and implemented an agent-based model. This solution allows us to deal with an heterogeneous population of (different types of) agents equipped with dynamic strategies and properties changing over time.

We are ultimately interested in better understanding the different strategies agents must be equipped with and the different social dynamics and relationships (e.g., different network topologies) allowing the system to reach specific equilibria.

We used information from the institutional website of Kickstarter to parametrize the model in order to maintain the specific ratio between the number of backers and the number of campaign promoters (i.e., 1/30) and to set the number of different categories of campaigns promoted.

We tested the scalability of the behavior of the model considering different group sizes, from few hundreds to thousands of agents.

ODD of the Model

The overview, design concepts, details (ODD) protocol is becoming more and more the standard way to present an agent-based model (ABM).

Proposed in 2006 (Grimm et al. 2006) to facilitate the understanding, transmission, and translation of any ABM, it gained the attention of modelers from different scientific fields, and it now represents the best way to present an ABM in a clear way allowing any potential reader to properly understand the model and to properly implement the model in all its details using any programming language/modeling platform.

Here, to present the model we developed, we refer to the last publication about ODD protocol standard (Grimm et al. 2010) and different sections useful to define and describe the model.

1. Purpose

   - *Question*: What is the purpose of the model?

- This model should clarify the dynamics of crowdfunding focusing on the specific dynamics of Kickstarter platform and all parties involved. The model studies three different types of agents: backers, campaign promoters, and Kickstarter. It investigates the complex, potentially non-linear dynamics emerging from interactions between campaign promoters (i.e., agents looking for resources to fund their campaign), backers (i.e., agents potentially funding other agents' campaign), and Kickstarter (i.e., the environment and the environmental conditions where behaviors and dynamics occur). We are interested in better understanding different strategies agents must be equipped with and different social dynamics and relationships (i.e., different network topologies) allowing the system to reach specific equilibria.

2. Entities, state variables, and scales

   - What kinds of entities are in the model? By what state variables, or attributes, are these entities characterized? What are the temporal and spatial resolutions and extents of the model?

   (i) *Agents/individuals*

      1. *Campaign promoters, backers, Kickstarter*

   (ii) *Spatial units* (e.g., *grid cells)*

      1. *Social space allowing interactions between agents (*i.e., *couple of agents playing the game)*

   (iii) *Environment*

      1. *All environmental conditions potentially affecting the outcome of agents' interactions*

   (iv) *Collectives*

      1. *Different types of agents are characterized by specific behaviors and properties:*

         (a) *Campaign promoters are defined by*

            (i) *A set of properties:*

               1. *Expected amount*
               2. *Ability in promoting the campaign*
               3. *Index of plausibility*
               4. *Tag of campaign*
               5. *Originality of the campaign*

            (ii) *And a set of behaviors:*

               1. *Strategies in planning a feasible campaign*
               2. *Strategies in implementing the campaign*

         (b) *Backers are defined by*

(i) *A set of properties:*

1. *A budget* (i.e., *the overall amount they can use to fund different campaigns)*
2. *Individual nonmonetary reward* (e.g., *how happy will the agent be if the specific campaign will reach the goal – this is the function of the actual amount reached by the specific campaign)*

(ii) *And a set of behaviors:*

1. *Strategies in choosing different campaigns to invest in*
2. *Maximizing own nonmonetary reward (distributions of tables of how appealing each different category is)*

(c) *Kickstarter is defined by*

(i) *Environmental features or properties:*

1. *Quality of information available to potential backers (from no information to complete information)*

(ii) *And behaviors:*

1. *Strategies in advertising different campaigns*
2. *Amount taken as fees (real Kickstarter takes 5%) + amount taken as payment processor fee (3–5%)*

3. Process overview and scheduling

- *Questions*: Who (i.e., what entity) does what, and in what order? When are state variables updated? How is time modeled, as discrete steps or as a continuum over which both continuous processes and discrete events can occur? Except for very simple schedules, one should use pseudo-code to describe the schedule in every detail, so that the model can be re-implemented from this code. Ideally, the pseudo-code corresponds fully to the actual code used in the program implementing the ABM.
- A Kickstarter configuration is initialized; campaign promoter agents are initialized; backer agents are initialized; all global variables (i.e., all other parameters different from the abovementioned types of agents) are initialized. The game is a multi-step and multi-run game. At step 1, run 1, backers are randomly coupled with a subset of promoters and decide how to allocate their budget to; agents have T rounds to play the first step of the game using the strategies they are equipped with (fixed strategies). Step 1, run 1, ends when backers played all rounds of the game. Step 2 is when campaign promoter agents implement campaigns or not and backer agents get (or get not) a reward afterwards. After Step 2 of the game, information about the state variables (e.g., the success rate of campaign promoters, the degree of satisfaction of backers) is saved. The final network configuration (i.e., the network topology – who interacted with whom) is saved.

4. Design concepts

   - Basic principles

   (i) Which general concepts, theories, hypotheses, or modeling approaches are underlying the model's design?
   (ii) The model gets inspiration from the well-established theories about profiling consumers' decision-making styles and aims to translate that approach into the context of crowdfunding phenomenon. We are interested in understanding cognitive mechanisms (and strategies) governing the behavior of campaign promoters and backers in the specific context of Kickstarter.

   - Emergence

   (i) What key results or outputs of the model are modeled as emerging from the adaptive traits, or behaviors, of individuals?
   (ii) Funds allocated to specific campaigns (in other words, funded projects, i.e., campaigns reaching at least the suggested goal amount) result as particular equilibria from the interaction between agents of three different types, those trying to get funds, those able to invest funds, and the platform "hosting" interactions and transactions. Moreover, specific social structures and preferential relationships (i.e., network topologies) emerge as results of the abovementioned interactions.

   - Adaptation

   (i) What adaptive traits do the individuals have? What rules do they have for making decisions or changing behavior in response to changes in themselves or their environment?
   (ii) The main adaptive trait agents have is the strategy (or the set of possible strategies) they adopt to play the two-step game. They can learn from their experience, and therefore they can modify their strategy in accordance with the outcome of past interactions.

   - Objectives

   (i) If adaptive traits explicitly act to increase some measure of the individual's success at meeting some objective, what exactly is that objective and how is it measured? When individuals make decisions by ranking alternatives, what criteria do they use?
   (ii) The main objective agents have depends on the group of origin: campaign promoter agents aim at reaching (at least) the goal amount they suggest at the very beginning of the "season"; backers aim at maximizing their non-monetary reward; Kickstarter aims at effectively advertising campaigns in order to get the fees. Therefore, campaign promoter agents act in order to get as much as possible from backer agents; backer agents act in order to allocate their budget to campaigns that succeed; and Kickstarter acts in order to maximize its fees.

- Learning

(i) Many individuals or agents (but also organizations and institutions) change their adaptive traits over time as a consequence of their experience. If so, how?
(ii) Agents modify their strategies as a consequence of the outcomes of past interactions: in other words, backer agents may decide to stop funding a specific campaign as soon as they perceive that the campaign is not being successful enough. Moreover, campaign promoter agents may change their strategies in order to attract more backers.

- Prediction

(i) Prediction is fundamental to successful decision-making; if an agent's adaptive traits or learning procedures are based on estimating future consequences of decisions, how do agents predict the future conditions (either environmental or internal) they will experience? If appropriate, what internal models are agents assumed to use to estimate future conditions or consequences of their decisions? What tacit or hidden predictions are implied in these internal model assumptions?
(ii) Backers allocate their budget to maximize nonmonetary reward based on learning mechanisms allowing them to adjust their investments over time.

- Sensing

(i) What internal and environmental state variables are individuals assumed to sense and consider in their decisions? What state variables of which other individuals and entities can an individual perceive, for example, signals that another individual may intentionally or unintentionally send?
(ii) We assume here that the information available to agents (either campaign promoter agents or backer agents) is a property of the environment, i.e., of the specific configuration of the current Kickstarter. This means that the amount and quality of information available to agents may vary from one scenario to another.

- Interaction

(i) What kinds of interactions among agents are assumed? Are there direct interactions in which individuals encounter and affect others, or are interactions indirect, e.g., via competition for a mediating resource? If the interactions involve communication, how are such communications represented?
(ii) We assume that the model models both direct and indirect interactions. Direct interactions result from the actual interaction between agents, more specifically combinations of one or more campaign promoter agent(s) and one or more backer agent(s). Indirect interactions result from the competition among campaign promoter agents in getting the resources from backer agents and from the competition among backer agents in getting the best nonmonetary reward from a funded campaign.

- Stochasticity

(i) What processes are modeled by assuming they are random or partly random? Is stochasticity used, for example, to reproduce variability in processes for which it is unimportant to model the actual causes of the variability? Is it used to cause model events or behaviors to occur with a specified frequency?
(ii) Stochasticity is used in different processes and to model all agents' properties. It is also partially used to implement strategies.

- Collectives

(i) Do the individuals form or belong to aggregations that affect, and are affected by, the individuals? Such collectives can be an important intermediate level of organization in an ABM; examples include social groups, fish schools and bird flocks, and human networks and organizations. How are collectives represented? Is a particular collective an emergent property of the individuals, such as a flock of birds that assembles as a result of individual behaviors, or is the collective simply a definition by the modeler, such as the set of individuals with certain properties, defined as a separate *kind* of entity with its own state variables and traits?
(ii) Collectives may affect the behavior of the system. Specific combination of strategies and properties of the different types of agents may affect the results. Particular groups (as groups of agents having same properties and behaviors) may emerge as consequence of interactions among agents.

- Observation

(i) What data are collected from the ABM for testing, understanding, and analyzing it, and how and when are they collected? Are all output data freely used, or are only certain data sampled and used, to imitate what can be observed in an empirical study ("Virtual Ecologist" approach; Zurell et al. 2010)?
(ii) Information about the state of variables (e.g., the success rate of campaign promoter agents; the degree of satisfaction of backer agents) is saved. The final network configuration (i.e., the network topology – who interacted with whom) is saved.

5. Initialization

- *Questions*: What is the initial state of the model world, i.e., at time $t = 0$ of a simulation run? In detail, how many entities of what type are there initially, and what are the exact values of their state variables (or how were they set stochastically)? Is initialization always the same, or is it allowed to vary among simulations? Are the initial values chosen arbitrarily or based on data? References to those data should be provided.
- A Kickstarter configuration is initialized; campaign promoter agents are initialized; backer agents are initialized; all global variables (i.e., all other

parameters different from the abovementioned types of agents) are initialized. All agents related variables are stochastically initialized. Numerosity of each type of agents (i.e., campaign promoter agents and backer agents) is defined.

6. Input data

   - *Question*: Does the model use input from external sources such as data files or other models to represent processes that change over time?
   - The model does not use any external input (e.g., configuration) file.

7. Submodels

   - *Questions*: What, in detail, are the submodels that represent the processes listed in "Process overview and scheduling"? What are the model parameters, their dimensions, and reference values? How were submodels designed or chosen, and how were they parameterized and then tested?
   - The model consists of a certain number of modules: campaign promoter agents' behaviors, backer agents behaviors, Kickstarter behavior, the two-step game played by interacting agents, the phase updating agents' state variables.

Table of Parameters, Ranges, and Initialization Values

Two Different Experimental Settings:

$N_{Backers} = \{200,400,800,1200,1600\}$	$N_{CampaignPromoters} = \{7,13,27,40,53\}$
$N_{Backers} = 800$	$N_{CampaignPromoters} = \{7,14,28,42,56\}$

Table 6.1 shows the parameters of the model together with range values and the values we used to initialize the model.

Profiling Investors

Inspired by the seminal work of Sproles and Kendall (1986) on profiling consumers' decision-making styles, we designed a bidimensional space where we can identify four main categories of backers' decision-making styles (see Fig. 6.1).

The two axes defining the space of different styles represent two fundamental features of all possible products a Kickstarter campaign may promote: novelty and reliability.

To profile backers' decision-making styles, we decided to range the relevance of these two features from low to high. Therefore, we can identify four possible extreme combinations of those values, notably:

1. Low novelty and low reliability

Table 6.1 Parameters of the model together with range values and the values

	Range	**Initial value (exp 1; exp 2)**
$N_{Backers}$	{200,1600}	[200,400,800,1200,1600] 800
$N_{CampaignPromoters}$	round($N_{Backers}$ / 30)	round($N_{Backers}$ / 30) [7,14,28,42,56]
$N_{Kickstarter}$	1	1
$id_{Backers}$	{1,$N_{Backers}$}	{1:$N_{Backers}$}
$id_{CampaignPromoters}$	{1,$N_{CampaignPromoters}$}	{1:$N_{CampaignPromoters}$}
$N_{Campaigns}$	{1,$N_{CampaignPromoters}$}	{1:$N_{CampaignPromoters}$}
Rounds	20	20
Pool.size		round($N_{CampaignPromoters}$ / 2)
Learning.interval		5
$N_{DifferentCampaigns}$		7
Max.budget		0.85
Max.expected. amount		2
Repetitions	{200,1000}	[200,1000]
Backers		
id	{1, $N_{Backers}$}	1:$N_{Backers}$
Learning	{0,1}	runif($N_{Backers}$,0,1)
Budget		maxBudget
Reward		0
preferredCampaign	{1,$N_{DifferentCampaigns}$}	sample(1:$N_{DifferentCampaigns}$, size = $N_{Backers}$, replace = T)
Marketing	{0,1}	runif($N_{Backers}$,0,1)
Cautious	{0,1}	runif($N_{Backers}$,0,1)
Eccentric	{0,1}	runif($N_{Backers}$,0,1)
Rational	{0,1}	runif($N_{Backers}$,0,1)
Stereotype	{1,4}	{1,2,3,4}
Campaigns promoters		
id	{1, $N_{CampaignPromoters}$}	1:$N_{CampaignPromoters}$
implementingAbility	{0,1}	runif($N_{CampaignPromoters}$,0,1)
expectedAmount	{0,Max.expected. amount}	sample(1:Max.expected.amount, size = $N_{CampaignPromoters}$, replace = T)
actualAmount	0	0
campaignTag	{1,$N_{DifferentCampaigns}$}	sample(1:$N_{DifferentCampaigns}$, size = x, replace = T)
abilityPromoting	{0,1}	runif($N_{DifferentCampaigns}$,0,1)
plausibilityIndex	{0,1}	runif($N_{DifferentCampaigns}$,0,1)
campaignOriginality	{0,1}	runif($N_{DifferentCampaigns}$,0,1)
campaignFeasibility	{0,1}	runif($N_{DifferentCampaigns}$,0,1)
Kickstarter		
infoQuality	{0,1}	runif($N_{Kickstarter}$,0,1)
advertisingStrategy	{0,1}	runif($N_{Kickstarter}$,0,1)
feesAmount	{0,0.05}	runif($N_{Kickstarter}$,0,0.05)

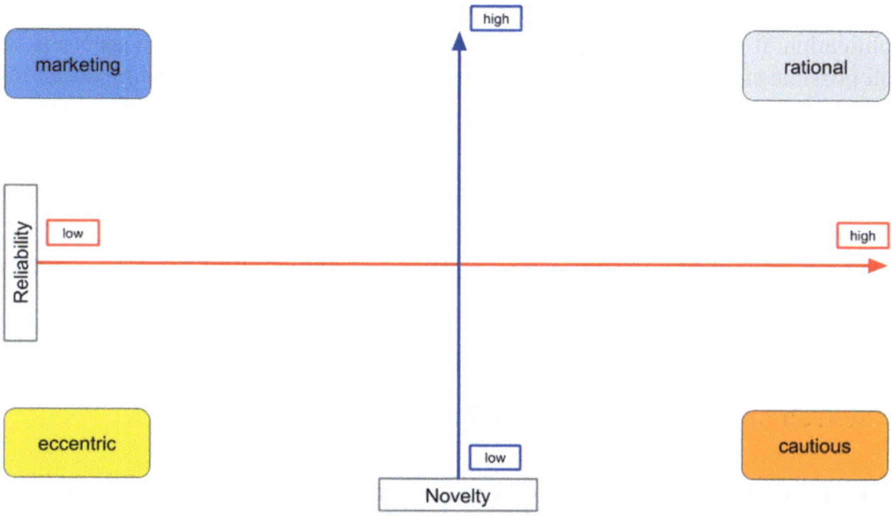

Fig. 6.1 Profiling investors. This figure shows the plane defined by "novelty" and "reliability" that we use to model four different strategies backer agents may be equipped with. Those strategies affect backer agents' decision-making in investing in different campaigns on Kickstarter

2. High novelty and low reliability
3. High novelty and high reliability
4. Low novelty and high reliability

At each of these extremes, we collocate a specific backers' decision-making strategy exemplifying a class of potential backers dealing with different campaigns promoted on Kickstarter.

These strategies are:

1. Eccentric
2. Marketing
3. Rational
4. Cautious

Eccentric backers are not interested in the novelty nor in the reliability of a product; their only concern is about the fact that they badly desire that specific product and no one else.

Marketing backers are "victims" of the market: this means that they aim to have the very last product, hoping to be the very first ones owning it.

Rational backers are economic agents in the very traditional acceptation: their utility function aims to maximize both the novelty and the reliability of the products they are interested in.

Cautious backers do not pay too much attention to the novelty of a product, because they are only interested in getting a product that is as much reliable as possible.

Of course, we are aware of the fact that this classification represents an oversimplification: it is not fully representative of the complex and variegated spectrum of all possible strategies backers may adopt in implementing their investing decision-making process. Nonetheless, we think that the abovementioned classification represents an interesting and useful roadmap to model backers' behaviors, and it captures the minimal requirements needed to properly model such behaviors in a heterogeneous population.

We use these four strategies to model the investing behavior of backer agents facing a series of (virtual) campaigns promoted on (virtual) Kickstarter by a certain number of promoter agents.

"Search and Invest": The Engine of the Simulator

The engine of the simulator is represented by a function implementing the decision-making process of backer agents. This function is used to model their strategies and the way those strategies affect their choice of investing a specific amount in a specific campaign.

From an algorithmic perspective, the model is a two-step model or game (in game theory terms).

In the first step, all backer agents have a certain number of rounds to invest their budget in different campaigns in accordance with their strategies and preference (i.e., the specific campaign label) and a learning parameter.

In the second step, campaign promoter agents can (or cannot) implement the campaigns they promoted.

More technically, the first step is implemented as a series of two nested "for" loops: the first loop is used to implement the number of rounds backer agents have to invest; the second loop is used to randomly select all backer agents to avoid any undesired effect caused by the specific order used to select them.

At each round, backer agents have access to a subset of campaigns (not all of them): this models a scenario where a real backer is unlikely to have access to information about all possible campaigns promoted on Kickstarter at the same time.

The result is that at each round a randomly selected backer agent is able to rank the subset of campaigns it has access to.

Finally, a learning mechanism is implemented in the model: at a specified time interval, backer agents' learning parameter is updated following a sigmoid function considering the current success of every campaign (see Eq. 6.1 for details).

$$l_i = \frac{1}{1+e^{-x}} \tag{6.1}$$

where

$$x = learningMatrix_{ij} + \left(\frac{actualAmount}{expectedAmount} * l_i \right)$$

with:

- i id of backer agent
- j id of campaign
- learningMatrix the matrix storing updated values of learning for any i backer agent and any j campaign
- actualAmount the funds campaign j received so far
- expectedAmount the campaign j goal
- learning the learning parameter of backer agent i

In the second step, campaigns are implemented (or not) depending on a particular property of campaign promoter agents modeling their ability in effectively implementing the campaign they promoted, assuming that they reached (at least) the target amount (i.e., the amount assigned at the very beginning of the simulation as the minimal amount allowing the campaign to be implemented – the "goal").

The "search and invest" function, the engine of the simulator, allows backer agents to rank all available campaigns in accordance with their strategy and preference.

Each backer agent is initialized with a specific campaign preference: this means that among all possible categories of campaigns (in Kickstarter each category represents a specific category of product, such as "games," "design," "technology," "film and video," "music," "fashion," "publishing," "food," "art," "comics," "theater," "photography," "crafts," "journalism," "dance"), a randomly selected different category is assigned to each backer agent.

This preferred category impacts on the investing decision-making of backer agents and on the reward they can receive at the end of the simulation: this happens if and only if at least one of the campaigns implemented is from the same category they prefer.

As soon as the backer agent ranked all accessible campaigns, it can invest a proportion of its budget available at that particular time (see Eq. 6.2 for details) in different campaigns following its own strategy (see section Profiling Investors for details) and preference (i.e., the category).

At the end of this two-step game, a series of information about backer agents, campaigns, campaign promoter agents, and interactions between different types of agents is saved.

$$budget_i = \frac{current\ budget_i}{number\ of\ remaining\ rounds_i} \tag{6.2}$$

Results

In this section, we report some results obtained running the model and present only a fraction of the different aspects of the system Kickstarter that can be investigated using our modeling approach.

Experimental Settings 1 This is the settings of the model for the following results.

Parameter	Value
$N_{Backers}$	200
$N_{CampaignPromoters}$	7
Replicates	10000

Success and Reward

Campaign's success rate: 0.3755
Rewarded backer's rate: 0.3198

The model we developed can be parametrized in different ways in order to mimic the behavior of Kickstarter under specific assumptions; moreover, it can be used to test theoretical hypotheses about strategies and dynamics underlying specific equilibria.

The very first result we present in this chapter concerns the campaign's success rate that a specific combination of the model parameters allows us to produce.

From the institutional website of Kickstarter, we can monitor the success rate of campaigns promoted so far. At the date we write (February 15, 2019), the overall success rate of Kickstarter considering all different categories as a single super-category is 36.66%.

The overall success rate produced by our model initialized with a specific combination of parameters (see experimental settings 1 on the table above) and averaging results over 10.000 replicates is 37.55%.

We would like to stress here that because of the strong stochasticity of initial conditions affecting every and any agent-based model, it is necessary to replicate the same model (i.e., the model with the same set of parameters) several times and to consider the results obtained averaging results over all replicates as the most reliable and meaningful ones.

Another interesting result of our model is the percentage of rewarded backers.

Considering the same combination of parameters producing the abovementioned results about success rate, we found that the proportion of backers receiving a reward after the implementation of the campaigns they invested in is 31.98%.

Since the genuine reward-based nature of Kickstarter (see Belleflamme et al. 2015 for details about reward-based systems), we think that it is potentially very interesting being able to investigate how the adoption of different strategies and how different environmental conditions may affect the proportion of backers receiving a form of reward.

An in-deep analysis of empirical data about rewarded backers in Kickstarter and the consequent comparison with our modeling results would be certainly of interest.

Descriptive Statistics and Statistical Analyses

In this section, we provide a simple descriptive statistics of the distribution of reward among backers.

Figure 6.2 shows the boxplot of the distribution of agents' reward. Table 6.2 shows the quartiles of such a distribution.

This table shows quartiles of the distribution of agents' reward on 10.000 replicates of the model, $\mathbf{N}_{\mathbf{Backers}} = 200$, $\mathbf{N}_{\mathbf{CampaignPromoters}} = 7$.

Results of reward obtained by backer agents is presented in Table 6.2 showing (from left to right) the extreme of the lower whisker, the lower "hinge," the median, the upper "hinge," and the extreme of the upper whisker, or in other words, the zeroth, first, second, third, and fourth quartiles of the distribution.

We use the chi-squared test to determine whether there is a significant difference between the expected frequencies and the observed frequencies of reward of agents belonging to the four subpopulations of agents adopting different specific strategy, namely, "marketing," "cautious," "eccentric," and "rational" agents.

Results show that there is no significant difference in frequencies of rewarded agents belonging to the four subpopulations (see Table 6.3).

Pearson's chi-squared test on M (Boolean matrix of reward frequency for different strategies)

This table shows the result of Pearson's chi-squared test on frequencies of rewarded agents from four different strategies.

We used analysis of variance to test whether the variance in reward can be explained by belonging to the abovementioned subpopulations of agents adopting specific strategies.

We did not find any statistically significant difference (see Table 6.4 for more details).

Anova:

$$y = \mathrm{lm}\left(Reward \sim stereotype\right).$$

This table shows the analysis of the variance (ANOVA) of reward in response to stereotype (which stands for one of the four subpopulations adopting different strategies).

Fig. 6.2 Distribution of reward. This figure shows the boxplot of reward obtained by agents. Data from 10.000 replicates of the model, $N_{Backers} = 200$, $N_{CampaignPromoters} = 7$

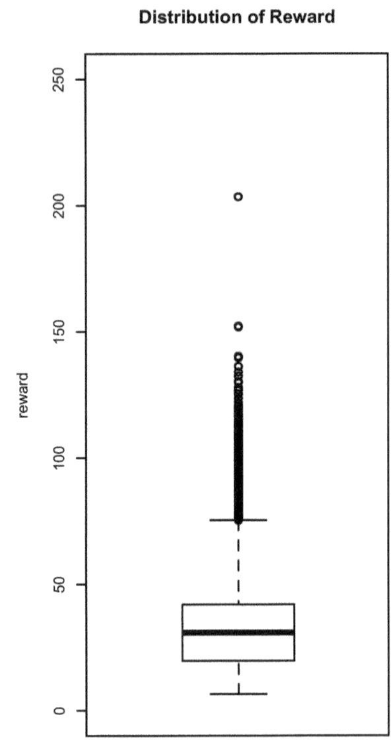

Table 6.2 Quartiles of the distribution of agents' reward on 10.000 replicates of the model, $\mathbf{N_{Backers}} = 200$, $\mathbf{N_{CampaignPromoters}} = 7$

0st	1st	2nd (i.e., median)	3rd	4th
6.4173	19.7098	30.8369	41.9642	75.3428

Table 6.3 The result of Pearson's chi-squared test on frequencies of rewarded agents from four different strategies

X-squared	df	p-value
1.3922	3	0.7074

Table 6.4 The analysis of the variance (ANOVA) of reward in response to stereotype (which stands for one of the four subpopulations adopting different strategies)

	Df	Sum Sq	Mean Sq	F value	Pr(>F)
as.factor(stereotype)	3e+00	332	110.83	0.3329	0.8016
Residuals	2e+06	665,891,014	332.95		

Results from chi-squared and ANOVA analyses do not surprise us: in fact, we think that in the real system (i.e., real Kickstarter) there is no direct causal connection between adopting a specific strategy and obtaining a specific reward, because ultimately reward of backers depends on the ability of campaign promoters to actually implement promoted campaigns.

It is not unusual observing a very well-invested Kickstarter campaign that is not actually implemented because one of many possible reasons (e.g., the promoter does not have the ability or the resources to produce on large scale the product(s) advertised on Kickstarter).

In other words, strategies adopted by backers do not impact on feasibility of campaigns.

Distributions of Investments

The profiling of different investing strategies produces a clear difference in the distributions of the amount invested by agents belonging to the four different above-mentioned subpopulations (Fig. 6.3 shows boxplots of distributions of invested amount).

On average, "marketing" and "cautious" agents invested less than "eccentric" and "rational" agents with "eccentric" agents investing more than any other type of agents (with a long tail of outliers).

Table 6.5 shows quartiles of the distributions of agents' investments for each different strategy.

This table shows quartiles of the distribution of agents' investment for each different strategy on 10.000 replicates of the model, $\mathbf{N}_{\mathbf{Backers}} = 200$, $\mathbf{N}_{\mathbf{CampaignPromoters}} = 7$.

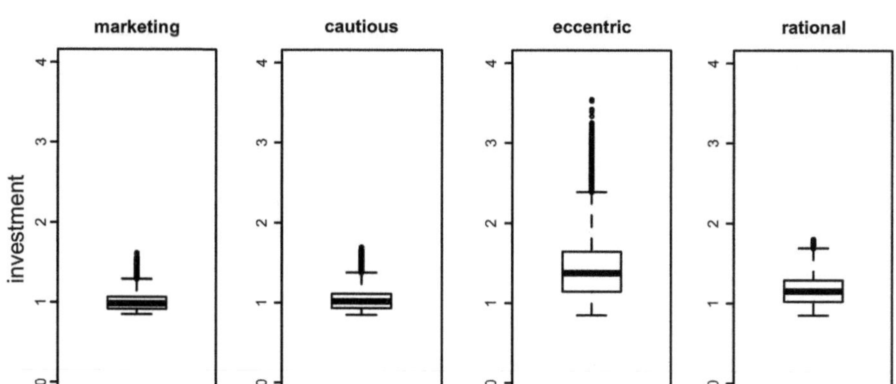

Fig. 6.3 This figure shows the boxplots of the distributions of "marketing," "cautious," "eccentric," and "rational" backer agents

Table 6.5 Quartiles of the distribution of agents' investment for each different strategy on 10.000 replicates of the model, $\mathbf{N}_{\mathbf{Backers}} = 200$, $\mathbf{N}_{\mathbf{CampaignPromoters}} = 7$

Strategy	0st	1st	2nd (i.e., median)	3rd	4th
Marketing	0.8500	0.9150	0.9828	1.0649	1.2897
Cautious	0.8500	0.9341	1.01848	1.1114	1.3774
Eccentric	0.8500	1.1437	1.3739	1.6419	2.3893
Rational	0.8500	1.0211	1.1495	1.2880	1.6883

Network Measures

Experimental Settings 2a This table refers to the parameters' values we used in this set of experiments.

Parameter	Value
$\mathbf{N}_{\mathbf{Backers}}$	{200, 400, 800, 1200, 1600}
$\mathbf{N}_{\mathbf{CampaignPromoters}}$	{7, 13, 27, 40, 53}
Replicates	200

We ran some experiments to investigate the effects of group size of backer agents and campaign promoter agents on the structure of the network resulting from agent-agent interactions.

We focused our analysis on two different measures.

The first measure is *centralization score* which is the whole graph-level centrality score based on node-level centrality measure (i.e., the *degree centrality* of nodes).

Historically, the first and conceptually the simplest among all different centrality measures available for analyzing social networks, the degree centrality of a node is defined as the number of links incident upon that node (i.e., the number of ties that a node has).

The degree centrality can be interpreted in terms of the potential of a node to have access to whatever is flowing through the network (such as information, resources, wealth, etc.).

In the case of a directed network (i.e., a network where links have a direction), we usually define two separate measures of degree centrality, namely, indegree and outdegree. Accordingly, indegree represents the count of the number of links directed to the node, and outdegree represents the count of the number of links that the node directs to others.

In sociology, usually, when links are associated to some positive aspects (e.g., friendship or collaboration), indegree is interpreted as a form of "popularity" and outdegree as a form of "gregariousness."

For the sake of simplicity, we assumed that in our model links are undirected, meaning that there is no qualitative difference between the link connecting agent A to agent B and the link connecting agent B to agent A.

Centralization (see Freeman 1979 for details) is a general method for calculating a graph-level *centrality score* based on node-level centrality measure.

The degree centrality of a vertex is its degree. The degree centralization of a network (or its centralization score CS) is the variation in the degrees of vertices divided by the maximum degree which is possible in the network of the same size.

$$CS_n = \frac{\Sigma(\max(x) - x)}{(length(x) - 1)} \tag{6.3}$$

with x = vector of all degree values in the network.

For degree measure, the most centralized structure is some version of the star graph, in-star, out-star, or undirected star.

Keeping constant the ratio between backer agents and campaign promoter agents (i.e., ratio = 30:1, as from "real" Kickstarter), result show that the centralization score of the networks resulting from the interactions between different type of agents decreases with the increase of the group size.

Figure 6.4 shows the boxplot of the distributions of the centralization score for different group size.

Table 6.6 shows the median values of these centralization score distributions on 200 replicates.

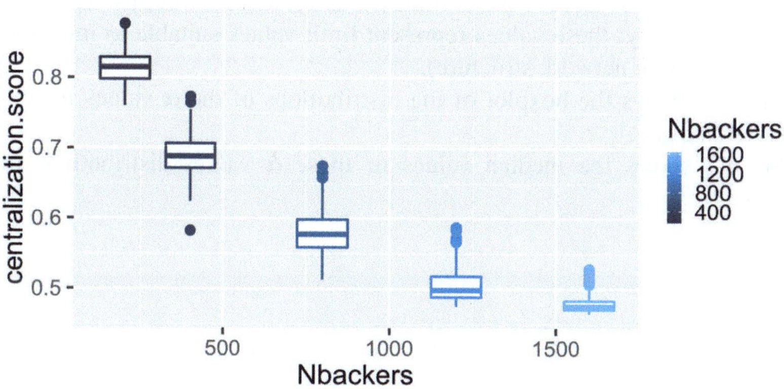

Fig. 6.4 Boxplot of centralization score distributions for different group size $N_{Backers}$ over 200 replicates

Table 6.6 Median values of centralization score distributions for different group size $N_{Backers}$ on 200 replicates

$N_{Backers}$ ($N_{CampaignPromoters}$)	200 (7)	400 (13)	800 (27)	1200 (40)	1600 (53)
Centralization Score (median)	*0.8155*	*0.6885*	*0.5775*	*0.4983*	*0.4752*

Median values of centralization score distributions for different group size $N_{Backers}$ on 200 replicates.

Another measure we decided to use to investigate the effect of group size on the structure of the interactions between agents is **α** which is the exponent of the fitted power-law distribution.

In other words, **α** tells us whether the network is a *scale-free network* or not.

A scale-free network is a network whose degree distribution follows a power law, at least asymptotically. That is, the fraction P(k) of nodes in the network having k connections to other nodes goes for large values of k as

$$P(k) \sim k^{-\alpha} \tag{6.4}$$

where **α** is a parameter whose value is typically in the range $2 < \alpha < 3$, although occasionally it may lie outside these bounds.

Also in this case values of **α** were calculated on 200 replicates, and therefore distributions refer to values obtained over 200 replicates of the same sets of parameters.

Results show that under the assumption of keeping constant the ratio between backers agents and campaign promoter agents (30:1 as from the empirical data retrieved from Kickstarter statistics webpage), networks resulting from the interactions between agents are not scale-free network in the majority of cases (except for the specific combination of $N_{Backers} = 400$ and $N_{CampaignPromoters} = 13$ and $N_{Backers} = 200$ and $N_{CampaignPromoters} = 7$ where the median values of **α** distributions are 3.0175 and 3.1074, respectively; these values represent limit values suitable to infer the presence of a scale-free network structure).

Figure 6.5 shows the boxplot of the distributions of the **α** values for different group size $N_{Backers}$.

Table 6.7 shows the median values of these **α** values distributions on 200 replicates.

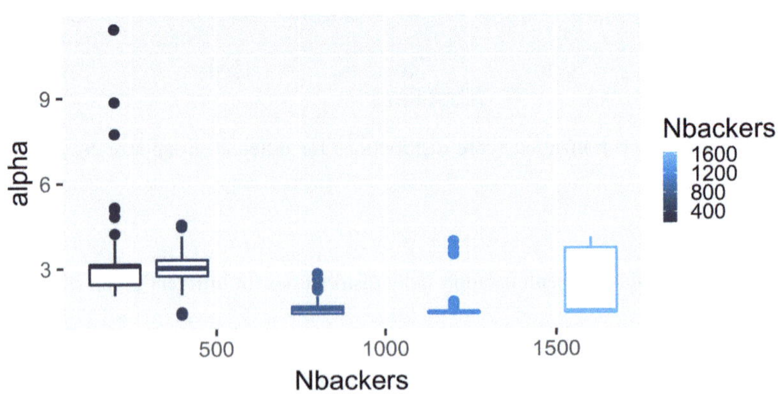

Fig. 6.5 Boxplot of **α** values distributions for different group size $N_{Backers}$ over 200 replicates

Table 6.7 Median values of $\boldsymbol{\alpha}$ values distributions for different group size $\mathbf{N_{Backers}}$ on 200 replicates

$N_{Backers}$ ($N_{CampaignPromoters}$)	200 (7)	400 (13)	800 (27)	1200 (40)	1600 (53)
Alpha (median)	*3.1074*	*3.0175*	*1.5703*	*1.4451*	*1.4638*

Median values of $\boldsymbol{\alpha}$ values distributions for different group size $\mathbf{N_{Backers}}$ on 200 replicates.

To test the robustness of our results and to further investigate the effects of group size together with different ratios between backer agents and campaign promoter agents, we decided to keep constant the number of backer agents and to vary only the number of campaign promoter agents in order to generate scenarios simulating conditions where the ratio between backer agents and campaign promoter agents is different from the ratio extrapolated from empirical data (i.e., from Kickstarter statistics website).

Experimental Settings 2b This table refers to the parameters' values we used in this set of experiments.

Parameter	Value
$\mathbf{N_{Backers}}$	800
$\mathbf{N_{CampaignPromoters}}$	{7, 14, 28, 42, 56}
Replicates	200

Results from these new sets of experiments show that centralization score is robust to changes in the ratio between the number of backer agents and the number of campaign promoter agents. The reason being obvious: the centralization score as explained above is a measure of the network referring to how centralized the network is; in other words, whether there are few nodes with a high degree or not. Because of the design of our model, those very connected nodes are always campaign promoter agents: these agents are the agents the most part of other agents (i.e., the backer agents) are linked with. Therefore, to make this network measure change, it is sufficient to modify the number of campaign promoter agents, and modifying the number of backer agents do not produce any significant effect.

Results in Fig. 6.6 shows exactly this: independently from the number of backer agents, the centralization score of a network with 7,14, 28,42, and 56 campaign promoter agents is qualitatively similar to the centralization score of the same networks with a different number of backer agents (compare results from Fig. 6.4 and results from Fig. 6.6).

Table 6.8 shows the median values of these centralization score distributions over 200 replicates.

Median values of centralization score values distributions for different group size $\mathbf{N_{CampaignPromoters}}$ on 200 replicates.

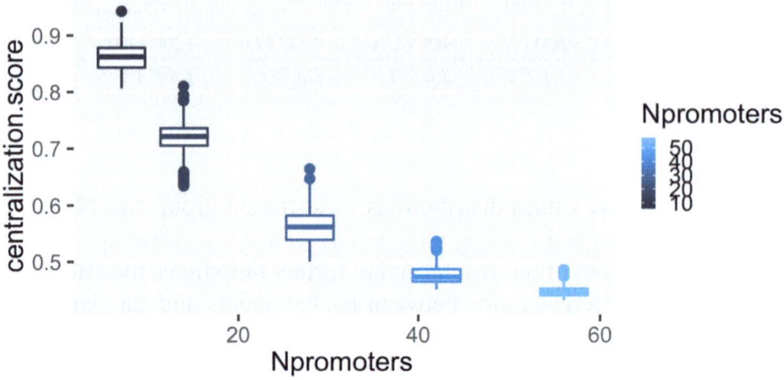

Fig. 6.6 Boxplot of centralization score distributions for different group size $N_{CampaignPromoters}$ over 200 replicates

Table 6.8 Median values of centralization score values distributions for different group size $\mathbf{N_{CampaignPromoters}}$ on 200 replicates

$N_{Backers}$ ($N_{CampaignPromoters}$)	200 (7)	400 (13)	800 (27)	1200 (40)	1600 (53)
Centralization Score (median)	*0.8627*	*0.7215*	*0.5623*	*0.4737*	*0.4468*

Results about **α** values distributions are consistent with results obtained in the previous sets of experiments.

In this case, **α** distributions obtained with the specific combination of $N_{Backers} = 800$ and $N_{CampaignPromoters} = 14$ and $N_{Backers} = 800$ and $N_{CampaignPromoters} = 56$ (Fig. 6.7) represent values strongly suggesting the presence of a scale-free network structure; in fact, median values of these two specific combinations are 2.8751 and 2.6643, respectively (see Table 6.9 for details).

Median values of **α** values distributions for different group size $\mathbf{N_{CampaignPromoters}}$ on 200 replicates.

Layout of a Network with Reward (i.e., with Implemented Campaign(s)) and Layout of a Network Without Reward (i.e., No Implemented Campaign)

Parameters and values	
Parameter	Value
N ($\mathbf{N_{Backers}} + \mathbf{N_{CampaignPromoters}}$)	207 (200 + 7)

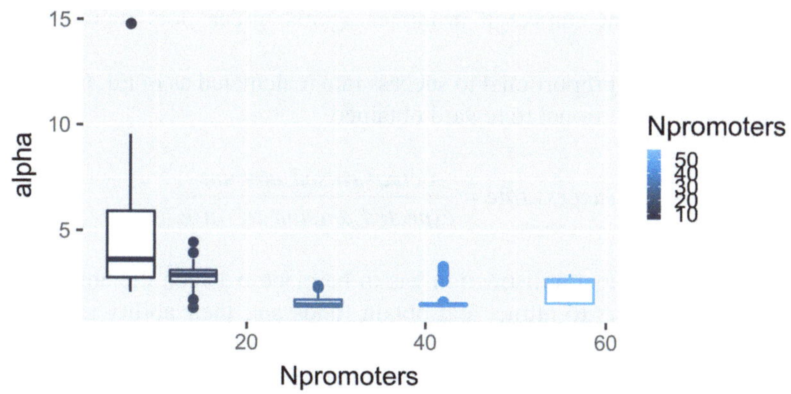

Fig. 6.7 Boxplot of α values distributions for different group size $N_{CampaignPromoters}$ over 200 replicates

Table 6.9 Median values of α values distributions for different group size $N_{CampaignPromoters}$ on 200 replicates

$N_{Backers}$ ($N_{CampaignPromoters}$)	800 (7)	800 (14)	800 (28)	800 (42)	800 (56)
Alpha (median)	*3.6076*	*2.8751*	*1.4854*	*1.5154*	*2.6643*

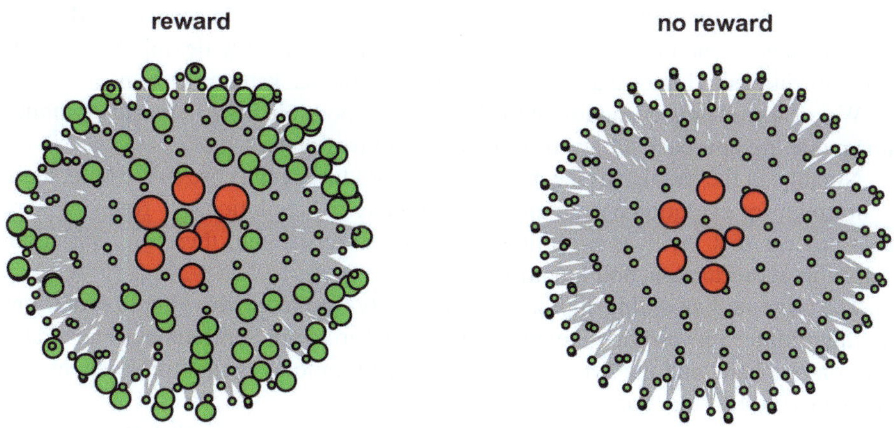

Fig. 6.8 Layout of networks of interactions of successful campaign(s) (left) and unsuccessful campaigns (right)

Figure 6.8 shows the different layout of a network where at least one of the campaigns has been implemented (on the left) and the layout of a network where no campaign has been implemented (on the right).

Red dots represent campaign promoter agents; green dots represent backer agents.

Size of red dots is proportional to success rate (calculated as in Eq. 6.5), and size of green dots is proportional to reward obtained.

$$success.rate = \frac{actual\ raised\ amount}{expected\ amount\ to\ raise} \quad (6.5)$$

Figure 6.8 shows that a distinction has to be made between the ability of campaign promoter agents to attract and obtain funds and their ability to implement promoted campaigns.

Network on the right shows a clear example of a system where campaign promoter agents have been able to attract and obtain funds (i.e., size of red nodes is changed), but not to implement campaigns: in fact, backer agents did not receive any reward (i.e., the size of green nodes is not changed).

Limitations of the Model and Future Directions

As every and any model is, the model we presented in this chapter represents a simplified version of a real system. Because of this and because of the arbitrariness of assumptions every and any modeling activity implies, our approach can be susceptible to criticisms and changes.

But, this is ultimately the way advancement of knowledge in science works.

Nonetheless, the model we presented in this chapter is very flexible and powerful, allowing us to investigate different aspects of the Kickstarter system.

We presented only a fraction of the broad set of possible behaviors and dynamics this modeling approach allows us to model and investigate.

Notably, in the settings of the model used to produce results presented here, there is no room for evolutionary dynamics (with the term "evolutionary" referring here to a process that involves evolution in biological terms and, therefore, variation, selection, reproduction, and mutation).

The presented modeling approach can be easily coupled with a simple evolutionary algorithm to investigate the evolution of backer agents' preferences and strategies on one hand and the evolution of campaign promoter agents' preferences and strategies on the other hand.

Moreover, adopting the same evolutionary approach, also Kickstarter behavior can be evolved.

These investigations may assume either the existence of pure strategies or mixed strategies (as per terms used in game theory and evolutionary game theory, respectively).

The agent-based modeling approach suggested in this chapter can also be used to perform what-if analysis: in other words, we could test some theoretical hypotheses about the effects of changes in some initial conditions.

For instance, in the models used to produce the results presented here, we always assumed that the success rate of campaigns promoted in Kickstarter is never 100%.

In the real Kickstarter system, there are different factors impacting on this element: ability of Kickstarter platform to promote the campaign; feasibility, plausibility, and originality of the campaign itself; quality of the information available on the Kickstarter platform to judge the campaign; and ability of the campaign promoters to implement the campaign once reached the target amount (the "goal" amount on Kickstarter webpage).

We implemented all of these factors as parameters of our model and assumed that each of them follows a particular distribution; all combinations of the possible values each of these parameters can assume allow us to model many different scenarios, and among all of these scenarios 100% of campaigns success rate represents only one of the many possible outcomes (a very unlike one).

But, what if we assume that there is a 100% campaigns success rate? In this scenario decision-making strategies and interactions dynamics may drastically and dramatically change.

Another very interesting aspect of Kickstarter our modeling approach allows us to investigate is the social structure emerging from interactions between different types of agents.

Social networks emerging from agent-agent interactions are used to "measure" the relationships resulting from different combinations of strategies, (environmental) conditions, and assortments.

This is a clear example of bottom-up approach where we leave the social structure to emerge and then we measure it.

But the opposite and complementary approach is possible, as well.

Using a top-down approach, we could impose a social structure (i.e., a particular network topology) from the very beginning and then observe the effects of the constraints this structure entails.

How campaigns success rate vary with varying the given network topology?

How investments distributions of different subpopulations of backers vary with varying the given network topology?

How Kickstarter strategies in promoting and managing campaigns vary with varying the given network topology?

One of the hot topics in economics is modeling economic behavior of (rational) agents and the evolution of their preferences.

Our modeling approach not only can be used to investigate this fascinating topic, but it can also be used to investigate the co-evolution of preferences (of different types of economic agents, e.g., backer agents and campaign promoter agents) and markets (e.g., a crowdfunding platforms as Kickstarter). The difference may seem not significant, but it actually has very relevant and profound implications both in terms of theories and applications.

Moreover, the model(s) we presented can be used to model the behavior of other stereotypes of (economic) agents from economic theory.

Finally, the model(s) we designed, developed, and implemented can be useful to investigate complex dynamics of economic markets and actors and to generate a massive amount of data (i.e., in silico generated data) that can be used to perform different traditional analyses, such as time series analyses and predictions about future trends of economic phenomena.

References

Belleflamme, P., T. Lambert, & A. Schwienbacher. (2010). *Crowdfunding: An industrial organization perspective*. Paper presented at Digital Business Models Conference: Understanding Strategies, Paris, France, June 25–26.

Belleflamme, P., Omrani, N., & Peitz, M. (2015). The economics of crowdfunding platforms. *Information Economics and Policy, 33*, 11–28. ISSN 0167-6245, https://doi.org/10.1016/j.infoecopol.2015.08.003.

Bouncken, R. B., Komorek, M., & Kraus, S. (2015). Crowdfunding: The current state of research. *International Business & Economics Research Journal (Online), 14*, 407.

Freeman, L. C. (1979). Centrality in social networks conceptual clarification. *Social Networks, 1*(3), 215–239.

Grimm, V., Berger, U., Bastiansen, F., Eliassen, S., Ginot, V., Giske, J., Goss-Custard, J., Grand, T., Heinz, S. K., Huse, G., Huth, A., Jepsen, J. U., Jørgensen, C., Mooij, W. M., Müller, B., Pe'er, G., Piou, C., Railsback, S. F., Robbins, A. M., Robbins, M. M., Rossmanith, E., Rüger, N., Strand, E., Souissi, S., Stillman, R. A., Vabø, R., Visser, U., & DeAngelis, D. L. (2006). A standard protocol for describing individual-based and agent-based models. *Ecological Modelling, 198*(1-2), 115–126. ISSN 0304-3800, https://doi.org/10.1016/j.ecolmodel.2006.04.023.

Grimm, V., Berger, U., DeAngelis, D. L., Polhill, J. G., Giske, J., & Railsback, S. F. (2010). The ODD protocol: A review and first update. *Ecological Modelling, 221*(23), 2760–2768. ISSN 0304-3800, https://doi.org/10.1016/j.ecolmodel.2010.08.019.

Hindriks, F., & Guala, F. (2015). Institutions, rules, and equilibria: A unified theory. *Journal of Institutional Economics, 11*(3), 459–480. https://doi.org/10.1017/S1744137414000496.

Kickstarter.com. Retrieved February 13, 2015. https://en.wikipedia.org/wiki/Kickstarter#cite_note-31.

Martin, T. (2012, April 12). *The jobs act of 2012: balancing fundamental securities law principles with the demands of the crowd*. Available at SSRN: https://ssrn.com/abstract=2040953 or https://doi.org/10.2139/ssrn.2040953.

Schwienbacher, A., & Larralde, B. (2010, September 28). Crowdfunding of Small Entrepreneurial Ventures. *Handbook of entrepreneurial finance*, Oxford University Press, Forthcoming. Available at SSRN: https://ssrn.com/abstract=1699183 or https://doi.org/10.2139/ssrn.1699183.

Sproles, G. B., & Kendall, E. L. (1986). A methodology for profiling consumers' decision-making styles. *Journal of Consumer Affairs, 20*(2), 267–279.

Zurell, D., Berger, U., Cabral, J. S., Jeltsch, F., Meynard, C. N., Münkemüller, T., Nehrbass, N., Pagel, J., Reineking, B., Schröder, B., & Grimm, V. (2010). The virtual ecologist approach: Simulating data and observers. *Oikos, 119*, 622–635. https://doi.org/10.1111/j.1600-0706.2009.18284.x.

Chapter 7
Fintech: The Recovery Activity for Non-performing Loans

Alessandro Barazzetti and Angela Di Iorio

Abstract In this chapter we present an algorithmic evaluation method for valuing loans secured by real estate. The model described represents the synthesis of the various components that qualify non-performing loans: the legal aspect, the real estate aspect, and the financial aspect. We have developed an algorithm capable of simulating the mechanism of the judicial auction, which relates the various quantities that qualify a credit, both in the supply phase and in the litigation phase. The model can be used to optimize a financial portfolio.

Keywords Non-performing loan · Judicial auction · Simulation · Real estate · Algorithm

What Are NPLs

Passing a non-performing loan (Non-performing loans: receivables for which there are more than 3 unpaid installments with notification to the debtor of default, n.d.) is the traumatic event that occurs when the debtor is no longer able to meet his debt: it denotes not only an economic loss for those who disbursed the credit but also highlights an error of judgment at the time of the rogation same. Certainly, the event of insolvency of a debtor is not always see predictable, even if you are in possession of all the necessary information, as they falls into the cases of so-called black swans (Taleb 2009).

However, even the credit in litigation or non-performing loan, according to the Anglo-Saxon definition (NPL), can represent a business opportunity: starting from

A. Barazzetti (✉)
QBT, sagl, Chiasso (Canton Ticino), Switzerland
e-mail: alessandro.barazzetti@qbt.ch

A. Di Iorio
DCM & Partners srl, Milan, Italy
e-mail: a.diiorio@dcmandpartners.com

© Springer Nature Switzerland AG 2019

F. Cecconi, M. Campennì (eds.), *Information and Communication Technologies (ICT) in Economic Modeling*, Computational Social Sciences,
https://doi.org/10.1007/978-3-030-22605-3_7

Table 7.1 The most important sales transactions in Italy in 2018

The greater transactions involving the sale of NPL in 2018				
	Seller	Million euros	Type	Buyer
05/2018	Mps	24.070,0	Npl secured e unsecured	Cartol. con Gacs, Italian Recovery Fund*
04/2018	Veneto Banca e Banca Pop. Vicenza	18.000,0	Npl e altri deteriorati	Sga
05/2018	Intesa Sanpaolo	10.800,0	Npl e altri deteriorati	Cartol. tramite spv Intrum-CarVal-Intesa SP**
12/2018	Banco Bpm	7.000-7.800	Npl (progetto Ace)	Cartol. con Gacs; Credito Fondiario ed Elliott*
06/2018	Banco Bpm	5.100,0	Npl secured (Project Exodus)	Cartolar. con Gacs
08/2018	Ubi Banca	2.750,0	Npl secured e unsecured	Cartol. con Gacs
12/2018	Società gestione crediti Delta (Sgcd)	2.100,0	Crediti in bonis e deterior. (portaf. Arcade)	Cerberus
12/2018	73 Bcc organizzate da Iccrea Banca	2.000,0	Npl	Cartol. con Gacs
11/2018	Bper	1.900,00	Npl secured e unsecrued (Progetto Aqui)	Cartol. con Gacs
06/2018	CreVal	1.600,00	Npl, soprattutto secured (Project Aragorn)	Cartol. con Gacs
11/2018	17 banche Popolari socie Luzzatti spa (di cui Banca Pop Bari cede 800 mln)*	1.600,00	Npl secured e unsecured Npl	Cartol. con Gacs (PopNpl srl)
05/2018	Sicilcassa	1.500,0	Npl	MB Finance (Gma srl)
07/2018	53 Bcc, casse rurali e banche popolari organizzate da Cassa Centrale Banca	1.397,0	Npl secured (39%) e unsecured (61%)	Varde Partners, Barclays Bank e Guber B.
05/2018	Fbs	1.280,0	Npl secured e unsecured	Banca Ifis (che compra anche Fbs)
09/2018	Unicredit	1.090,0	Npl unsecured (progetto Torino)	Banca Ifis
01/2018	Bnl Bnp Paribas	1.000,0	Npl unsecured retail (portafoglio Saturnia)	Alcudi spv srl
06/2018	Banco Desio	1.000,0	Npl secured e unsecured	cartol con Gacs
07/2018	21 Bcc organizzate da Iccrea Banca	1.000,0	Npl	cartol con Gacs
06/2018	Banco di Sardegna (gruppo Bper)	900,0	Npl soprattutto secured	Cartol con Gacs - 4Mori Sardegna srl
08/2018	Cariparma	700,0	Npl soprattutto secured	Pimco
07/2018	Biverbanca e CR Asti	697,0	Npl secured	cartol con Gacs
11/2018	Unicredit	675,0	Npl secured	Fortress
09/2018	Banca Intermobiliare	600,0	Npl	Cart. Nuova Frontiera - Attestor Val. Master F.
07/2018	Unicredit	537,0	Npl secured e unsecured	Banca Ifis
07/2018	Crédi Agricole (Cariparma)	450,0	Utp (Project Valery)	Bain Capital Credit
09/2018	Crc Bayview (mercato secondario)	425,0	Npl unsecured personali	MBCredit Solutions
11/2018	Unicredit	384,0	Npl unsecured corporate	J-Invest

Source: Milano Finanza on public data processing carried out by BeBeez (2019)

the 1990s and thanks to the law on securitization (Leggen 1999), which defined a precise regulatory framework, NPL trading operations reached high trading volumes (PWC 2019). The following Table 7.1 shows the transfers made during 2018, the name of the seller and the buyer with the different types of credit negotiated, and the relative volumes.

Some of these loan portfolios have been securitized through financial vehicles or special purpose vehicles (SPVs), set up ad hoc, with the following issue of securities: currently the SPVs in Italy are 937 (BdI 2019).

The importance of sales and securitization transactions consists in the possibility of transferring price and credit risks and generating new liquidity, new credit capital, and additional own resources (Porzio 2001).

They therefore represent a tool available to credit institutions to reduce the effects of losses on the financial statements due to the increase in the number of non-performing loans.

In recent years we have witnessed a slowdown in this activity, due to an overvaluation of portfolios in the financial statements of the transferring institutions, which have caused the sale prices to levitate and which have therefore become inadequate for those who buy (the buyer is instead linked to the negative real estate

Table 7.2 Non-performing positions of the main Italian banking institutions at 4 q 2018

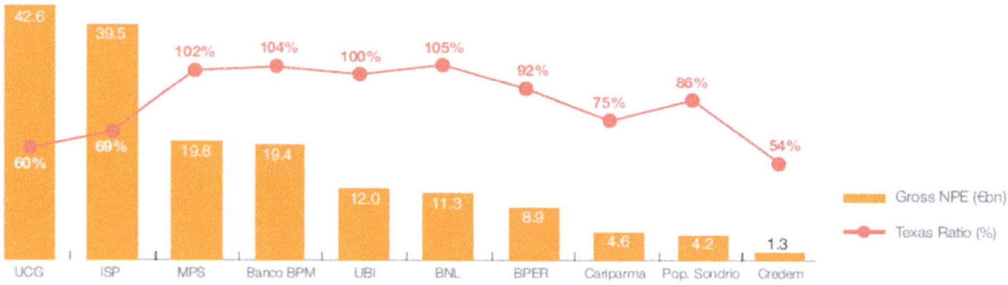

Source: PWC (2018)

dynamics of the assets pledged as collateral). However, the strong accumulation of problematic positions at banking institutions suggests that the issue of disposals, management outsourcing, and securitization will inevitably arise again.

If we analyze the data in the reports of the banks that published quarterly reports (8 banking groups including the 5 big ones that are included in international stress tests: Unicredit, Intesa, MPS, Banco Popolare and UBI, plus BNL, BPER, Cariparma and Pop.Sondrio) it is noted that the total of gross impaired loans or non-performing loans (NPE) rose to over 162 billion euros, suggesting the continuous sales of big portfolios of loans (Table 7.2).

Due Diligence and NPL Management Activities

The correct due diligence, or the activity of analysis of the credits carried out by the buyer on data provided by the credit seller, is therefore an important topic in the financial context, and it is in this framework that we propose our study on the relationship between the supply conditions of the credit and the conditions for its recovery if in dispute.

The fundamental themes of the due diligence in the NPL purchase phase are the selection of credits and the analysis of recoverability. The study of the mechanisms for the transfer of non-performing loans and their method of recovery can also help us to understand if there are different rules of engagement on the part of those who, by disbursing the loan, hold the action strategy for its recovery: it can decide for an out-of-court settlement or for the a priori selection of credits that should be brought back to performing status with respect to those that it is better to sell to third parties, fixing the price with greater precision.

A credit of banking origin, guaranteed by a real estate mortgage, is said to pass to suffering (In Italy the definition of problem loans is provided by the Bank of Italy, 2009) whenever the debtor does not honor the debt contracted with the creditor (Alessandrini 2011). The creditor then exercises his right to recover the residual debt, plus the related interest calculated from the date of the last payment, through a court action: a credit dispute then arises.

The activity of recovering the residual credit consists therefore in placing the immovable property in auction through the start of a legal action in the court of territorial jurisdiction.

In our analysis we take into consideration the subset of the credits for first-degree mortgage loans and real estate for residential use or the "first-home" mortgages, as they represent the most common and widespread type of loans passed to bad debts.

We ask the questions if there is an algorithm able to predict the recoverability of non-performing loans.

The rest of the work is therefore organized as follows: we will describe in detail how it is possible to create a representative model of the logical structure of the variables that describe a credit passed to non-performing. By assigning numerical values to the variables that are consistent with reality, we will calculate credit recovery forecasts and related timing.

Fintech Application: An Algorithmic Solution

An Algorithmic Evaluation Method for Valuing Loans Secured by Real Estate

To give both a quantitative and qualitative answer to the previous question, we have developed an algorithm capable of simulating the mechanism of the judicial auction, which relates the various quantities that qualify a credit, both in the supply phase and in the litigation phase. The algorithm allows to calculate the award amount of a judicial auction and the timing for each phase from the start of the legal action to the final allotment, note the initial date, and calculate the date of actual allotment of the amounts recovered.

In order to define a simulation model of the judicial auction mechanism, we must first define the conceptual architecture of the NPL: Fig. 7.1 shows what we call "evaluation ring" or the related set of entities that make up the noncredit mechanism performing.

In this diagram the debtor is the subject of suffering: let's start from the debtor and analyze the graph counterclockwise. The counterclockwise route is also called the "static cycle" and concerns all those entities that define the credit in dispute; these quantities are constant over time (in this sense the concept of static is defined) or in any case subject only to variations in amounts, based on predetermined rates (e.g., interest on late payments increases by a fixed annual percentage). The static cycle is therefore associated with all the amounts and dates that characterize the credit both in the disbursement phase and in the actual dispute: these values include the amount disbursed and the related disbursement date, the amount passed to non-performing, and the relative date and amount for the expiration of the benefit of the term and the relative date. Always following the ring in a counterclockwise direction, we meet the mortgage, which is also constant over time both in terms of amount and degree: it is a burden on the building, set up to guarantee the credit disbursed.

Fig. 7.1 Evaluation ring

Continuing along the ring, we find the building: the data that characterize it are its value at the time of the supply (the OMV), its current value restated through an appraisal of updating, its intended use (residential, commercial, etc.), and its typology (apartment, villa, office, etc.).

Let us now return to the debtor and walk the ring in a clockwise direction: the path is now called the "dynamic cycle," to better highlight the fact that the entities that characterize it change over time as they are subject to the intervention of third parties, namely, the court, which independently and unquestionably manages the steps that lead to credit recovery. The first entity we encounter is the legal procedure, which is activated for the recovery of the credit, be it a real estate attachment or an insolvency procedure, in the case of a company (Giusti 2012). The legal action is activated by the creditor through a legal act in the court, following which the court itself notifies the debtor to initiate the legal procedure against him. The court then creates an auction lot, which includes the property guarantee: we then close the ring and return to the property.

Each of the entities that make up the ring is linked to the others through a "many-to-many" link: a debtor can have multiple credits with different mortgages on different properties; only some of these credits can be in dispute, and therefore only some of these properties will become auction lots of different legal procedures. The graph represents an opportune simplification of the number of subjects and entities involved in the credit dispute: we do not consider out-of-court solutions or the compositions out of court of the dispute as we do not consider the presence in the suffering of third parties but jointly and severally liable credit (guarantors, third-party mortgage lenders, etc., which in turn can be objects of legal action for the recovery of credit, not only by mortgage but also by unsecured credit). The reason for this simplification work is dictated by the nature of what we want to prove, which is not a priori linked to other entities, other than those of the evaluation ring as illustrated in its simplified version.

Algorithm Processing and Operation

We now build our model on the basis of the evaluation ring architecture and work out the algorithm that calculates the award value by simulating the judicial auction process and finally determining how much the lender gets and how quickly (Bertuglia and Vaio 2011). Let's start by defining all the parameters of the model and its operation following the detail in the flowchart of Fig. 7.2.

The starting point of the elaboration is therefore the real estate value that we call OMV (open market value) that is the value of the property on the free market. It is the value obtained from the real estate appraisal, which is carried out at the time of the loan. In our model we hypothesize to deal with mortgage loans of a land and first-degree nature: therefore, we set the mortgage value as a percentage of the OMV.

Let us now consider the amount that is passed to bad debts as a percentage of the credit disbursed, and we call this value GBV (gross book value). In this way we will be able to model different situations of passage in litigation, that is, situations in which suffering is immediate or with residual debts even very low. The GBV is a value that takes into account interest and expenses accrued in the period between the transition to non-performing and the start of the legal procedure, as calculated by the creditor, but it will only be the court that can process the correct counts of the interests of late payments calculated from the date of transfer to non-performing. We call this amount DBT and establish that it is a percentage that is always less than 100% of the GBV, according to a consolidated historical trend. The amount of DBT will then be charged with the additional legal interests up to the time of the award according to what established by the law ex 2855 cpc.

The process of the auction mechanism envisages, in its initial phase, that the creditor proceeds with a legal procedure for the recovery of the residual debt not

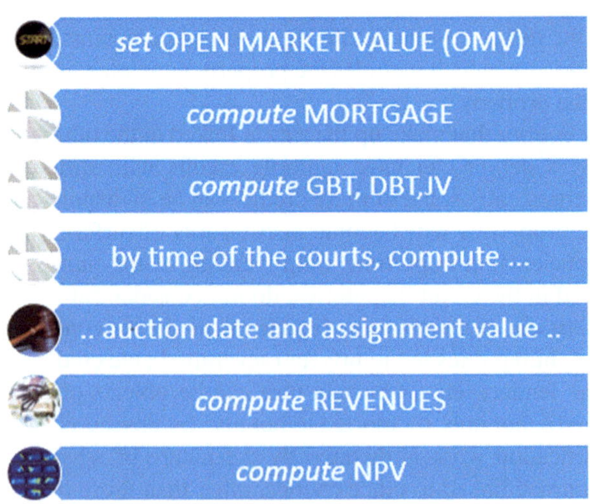

Fig. 7.2 Flowchart of the algorithm

collected by the court of jurisdiction for geographical location of the asset to guarantee the credit itself. The court, appointed the technical expert or CTU, will proceed to the composition of the sale lot of the real estate and, having established the minimum value, will put it at auction at that base value. If the judicial auction is to be deserted, the judge will set a new auction by reducing the value of the lot by 25%, according to the current legislation. In our model we consider the auctions in succession from the first to the nth adjudication: we assume that the auction without enchantment, foreseen by the legislation that came into force in 2006, is a first auction with a negative outcome (The code of civil procedure Articles from 483 to 604 2006).

How can we simulate the auction award process? We therefore ask ourselves how we can determine, by means of a calculation, which auction will be awarded a property. We determine the sale auction by putting in competition the real estate value given by the court, or by the court expert or by the appraisal to the disbursement, with the "real" real estate value of the property that is the value to which a speculator takes part in the auction to buy estate. It takes the name of judicial real estate value or JV (judicial value) and is a value that in the activity of credit management is usually determined by property technicians experienced in the NPL market (The Italian Banking Association (ABI) 2017). This type of evaluation does not only take into account the value of the property on the free market but also takes into account the "speculative" aspects of intervention in the real estate auction and therefore is also based on an in-depth examination of the area in which the property is located, property subject to judicial auction and its possible real estate developments. Furthermore, this evaluation also takes into account the type of debtor, the circumstance that the debtor is occupying the property, or that it is rented to third parties, of how much it will cost to clear it and of how long it will take, of the debtor's belonging to particular categories social, etc. In our model we assume that JV is a percentage higher or lower than 100% of the OMV value. If the supply appraisal was correct, then the value of JV will not deviate much from the value of OMV; on the contrary, if the real estate valuation at the supply was wrong, then we must expect a large volatility of the value of JV compared to the OMV. The assumption we make is that if a bad real estate valuation has been made to the disbursement of the mortgage, the JV will result in a value that will be, with high probability, extremely different from the OMV. On the other hand, a "good" payment will lead to a higher probability that, even when the dispute is underway, the JV real estate value will be in line with the original OMV value.

At this point we calculate the sale auction according to the formula (7.1):

$$n = \frac{\log\left(\dfrac{JV}{OMV^{*} * (1-T)}\right)}{\log(1-rilancio)} \tag{7.1}$$

where

n is the award auction

JV is the real estate value as per Table 7.5

$$\text{OMV}^* = \text{OMV} * \left(1 + \text{perc\_ril}\right) \tag{7.2}$$

where

perc_ril is the re-launch and OMV is a real estate value as per Table 7.5
T is the tolerance as per Table 7.5

Another fundamental fact in the simulation model is the timing of the courts. For each legal phase, we set a different duration expressed in months. We also consider the only legal phases of real estate attachment, which is the legal action that we expect to be undertaken for the type of receivables considered or disbursement by home mortgage to an individual (see Table 7.3).

The sum of the durations of the individual legal phases, from the deposit of hypocatastal documentation to the adjudication auction and therefore to the actual distribution of the proceeds, constitutes the timing of the court. The timing of the individual courts is provided by the T6 Association study (The study is freely downloadable at the web address n.d.) (T6 Association, Study of court times 2017, 2018).

Once the auction number has been set, and starting from today's date, the timing of the court is known, we then determine the award date. Once the sale auction has been determined, through the auction discounts mechanism, we can establish the amount by which the property must be awarded starting from its OMV value (e.g., if the property is awarded at the second auction, the amount award procedure will be OMV $*$ 0.75%). Abatements continue until the lot is awarded: at this point the judge will allocate the proceeds of the sale among the creditors according to the priority established by the mortgage degrees.

From the creditor's point of view, we can quantify the proceeds from the sale of the asset at auction as the minimum between the award value, the value of the mortgage, and the credit claimed. At this point we establish the recovery amount; gross of the expenses incurred, such as the minimum value between the DBT amount; increased legal interest from the date of today to the award date; the mortgage amount; and the award value. This value is called the gross creditor or RLC revenue:

$$\text{RLC} = \min\left\{\text{award value}, \text{DBT} + \text{interst}, \text{mortgage amount}\right\} \tag{7.3}$$

Table 7.3 The legal phases of the proceedings

Description
Start of the legal procedure: search for land registry documentation
CTU appointment
CTU deposit
Sale auction 1 Sale auction n
Auction revenue or distribution of sums

Table 7.4 Legal phases and legal and procedural costs

Description	Expense
Search for land registry documentation	€ 4.000
CTU appointment	€ 250
CTU deposit	€ 2700
Auction sale 1..n	€ 1000
Auction revenue or distribution of sums	€ 4300

Regarding expenses in Table 7.4, we indicate those taken into consideration for a standard real estate attachment procedure, and associated with each phase, we also consider the related costs expressed in euros.

We now determine the legal expenses, according to the standard rate table in Table 7.4, for the number n of auctions, and subtract them from the gross creditor revenue, obtaining the net income of the creditor or RNC:

$$RNC = RLC - Expenses \tag{7.4}$$

This amount must then be placed over time on the date of distribution of the auction revenue. The last step therefore consists in calculating the value updated to date NPV (net present value) (Lee and Lee 2010) of the RNC, according to a conventional rate, which in the whole simulation is set at 10% to have a common basis of comparison.

The net creditor revenue (RNC) and the discounted amount (NPV) represent the two parameters with which we judge the "quality" of the dispute or the creditor's ability to recover his credit passed to non-performing and in dispute. While the RNC gives us an absolute indication of how much is recovered, the value of NPV allows us to assess how much is collected in relation also to when it is collected, once the expected return has been set. As benchmarks we consider the percentages that are obtained by making the RNC/GBV and NPV/GBV ratios: in this way we can correctly compare the results obtained in the two cases of "good" and "bad" delivery. This allows us to assess the ability to recover credit in litigation without knowing qualitative aspects that cannot be modeled. We recall, in fact, that we are talking about a simulation model that leads to the determination of a collection forecast based only on quantifiable and logically variable parameters: we do not simulate what cannot be quantified reasonably.

We define the parameters that intervene in the model in Table 7.5: the parameters are subdivided according to their origin between endogenous values or characteristic of the single credit and exogenous values of environmental nature and referable therefore to the condition in which the credit is found.

They are also divided into areas of belonging with respect to the evaluation ring.

The model also uses parameters, which we define as secondary, whose task is to better adapt the simulation to reality. We therefore introduce a tolerance index T that sets a minimum deviation between the value of JV and OMV below which the

Table 7.5 Description of the parameters

Parameter	Description	Area of belonging	Parameter origin
OMV	Property value at free, coincides with the value assigned by the court (CTU)	Building	Endogenous
JV	True value of the property determined by the real estate specialist when the claim is already in dispute	Building	Endogenous
Decline	Rate of auction discount (25% new legislation)	Legal	Exogenous
T	Tolerance percentage in the deviation between the OMV and JV value	Legal	Exogenous
perc_ril	Re-launch percentage during the auction	Legal	Exogenous
TIMING	Time in months for the court to move from one legal phase to another	Legal	Exogenous
Year interests	Default interest maturing on the residual debt	Legal	Exogenous
Percent GBV on used	Percentage by which we determine the residual amount of credit (GBV) with respect to what was disbursed	Credit	Endogenous
Percent delivered on OMV	Percentage of the credit that is paid compared to the real estate value	Credit	Endogenous
Mortgage value	Mortgage amount	Credit	Endogenous
DBT	Forfeiture of the benefit of the term or counting of the actual residual credit	Credit	Endogenous

immovable property is anyway awarded at the current auction without abatement. This parameter allows us to avoid passing from one auction to the next when, as reasonably happens in reality, the value proposed by the court can be considered close to what the speculator expects, who intervenes at the current auction without waiting for the subsequent slaughter. We also introduce a bid for each auction, equal to a percentage of the amount of the current auction base, to simulate the competition in the auction. As it happens in reality, we therefore expect that there will be more speculators in the auction and therefore at least a rise.

Valuation Examples

Now let's try to elaborate the model setting the minimum and maximum values of the variables and impose a random choice of the values included in this range as shown in Table 7.6.

After setting the values we then elaborate and report the model for five simulations as shown in Table 7.7, with the results of the model.

Table 7.6 Parameters used to generate the numbers in Table 7.4

Parameter	Distribution type	Min	Max
OMV	Random number between min and max	100,000.00	600,000.00
JV	Random number between min and max	60%	140%
Ribasso	Constant	25%	
T	Constant	5.0%	
perc_ril	Constant	3.0%	
TIMING	Random number between min and max	4	12
Year interests	Constant	1.0%	
Percent GBV on used	Random number between min and max	40%	70%
Percent delivered on OMV	Random number between min and max	80%	120%
Mortgage value	Random number between min and max	90%	110%
DBT	Random number between min and max	70%	90%

Table 7.7 Some simulations

# Credit	1	2	3	4	5
Amount disbursed	333,990.92	238,734.96	458,687.97	397,908.87	265,823.75
Amount of GBV	210,414.28	155,177.72	215,583.35	167,121.73	116,962.45
Mortgage amount	403,232.94	193,457.64	360,677.72	406,466.05	248,752.50
DBT	180,956.28	116,383.29	183,245.84	143,724.68	100,587.71
DBT date	08/31/2017	08/31/2017	08/31/2017	08/31/2017	08/31/2017
DBT ex2855	186,384.97	119,874.79	190,575.68	148,036.42	104,611.22
OMV = CTU	407,306.00	205,806.00	392,041.00	427,859.00	243,875.00
JV	250,840.26	118,154.36	252,836.90	236,166.83	109,487.50
Awarding amount	233,691.82	118,081.19	224,933.52	245,484.10	104,942.46
Award month	34	34	34	34	38
Award date	01/01/2020	01/01/2020	01/01/2020	01/01/2021	05/01/2021
RLC	195,326.68	118,081.19	197,616.24	158,095.08	104,942.46
Expenses	−14,370	−14,370	−14,370	−14,370	−15,455
RNC	180,956.28	103,710.79	183,245.84	143,724.68	89487.51
NPV	135,524.11	77237.60	137,251.75	107,430.57	64020.22
RNC / GBV	86.00%	66.83%	85.00%	86.00%	76.51%
NPV / GBV	64.41%	49.77%	63.67%	64.28%	54.74%

Conclusions

The model described represents the synthesis of the various components that qualify non-performing loans: the legal, real estate and financial aspects.

The fundamental parameters lie in fact in the correct determination of the census of the legal situation (legal action in progress, legal phase completed and forthcoming, information on the CTU value deposited, and possibly the price of the current auction base) and the technical aspects that allow the identification of the property placed as a guarantee and related credit and accounting situation at the time of transfer to non-performing.

Added to this is the determination of the correct value of the property updated to date on the free market (OMV) and the judicial value (JV) of the building.

Another fundamental aspect is the knowledge of court times to allow the allocation of cash flows expected over time.

The model therefore lends itself to the development of the portfolio of doubtful loans with the realization of business plans that predict the cash flows deriving from the judicial recovery of the individual property placed as a guarantee.

References

Alessandrini, P. (2011). Economia e Politica della moneta, Il mulino.
BeBeez. (2019). Milano Finanza on public data processing carried out by BeBeez.
Bertuglia, C. S., & Vaio, F. (2011). Complessità e modelli, Bollati Boringhieri.
BdI. (2019). Banca d'Italia: elenco delle società veicolo di cartolarizzazione (SPV).
Giusti, G. (2012). Il pignoramento immobiliare, Giuffrè.
Non-performing loans: receivables for which there are more than 3 unpaid installments with notification to the debtor of default.
Lee, C. F., & Lee, A. C. (2010). *Handbook of quantitative finance and risk management* (Vol. 1). Springer-Verlag.
Leggen. 130 of 30 April 1999 containing "Provisions on the securitization of receivables".
In Italy the definition of problem loans is provided by the Bank of Italy in circular letter n. 272 of 30 July 2008 concerning banking and financial supervision regulations Circular no. 272 of 30 July 2008 – 1st update of 10 December 2009.
Porzio, C. (2001). Securitisation e crediti in sofferenza, Bancaria Editrice.
PWC. (2018). The Italian NPL market.
PWC. (2019). The Italian NPL market.
Taleb, N. (2009). Il cigno nero. Come l'improbabile governa la nostra vita, Il Saggiatore Tascab.
The code of civil procedure Articles from 483 to 604 of the code of civil procedure (updated to the law of 24 February 2006, number 52) Book III – of the Execution Process Title II – Of forced expropriation.
The Italian Banking Association (ABI) in 2017 issued the guidelines for the correct determination of the value of Judicial Value.
The study is freely downloadable at the web address.: www.osservatoriot6.it

Chapter 8
CDS Manager: An Educational Tool for Credit Derivative Market

Federico Cecconi and Alessandro Barazzetti

Abstract How could we teach the dynamics of new financial instruments, not very much understood and potentially dangerous? In this work we show some limitations of classical financial learning framework, which uses a description of the dynamics through mathematical models, and we propose an alternative approach, based on the agent-based modeling framework. In the model presented, an operator observes, through a simulation, the behavior of other artificial agents in its own market (Liu et al., J Manag Inf Syst 7(1):101–122, 1990; Radicchi et al., Proc Natl Acad Sci USA 101(9):2658–2663, 2004; Yang et al., Int J Electron Commer 6(1):101–102, 2001).

Keywords Credit derivative market · Agent-based modeling · Credit default swap · Networks · Educational tools

How you can teach the dynamics of new financial instruments, not much understood and potentially dangerous? In this work (i), we show some limitations of classical financial learning framework, which uses a description of the dynamics through mathematical models, and (ii) we propose an alternative approach, organized on a computational model based on agents. In our model, an operator observes, through a simulation, the behavior of other agents (artificial) in its own market, and he is able to assess how the behavior of artificial agents *dirty* the significance of variables on which is taking its decisions, for example, the price.

F. Cecconi (✉)
LABSS ISTC-CNR, ROME, Italy
e-mail: federico.cecconi@istc.cnr.it

A. Barazzetti
QBT Sagl, Chiasso, Switzerland
e-mail: info@qbt.ch

© Springer Nature Switzerland AG 2019
F. Cecconi, M. Campennì (eds.), *Information and Communication Technologies (ICT) in Economic Modeling*, Computational Social Sciences,
https://doi.org/10.1007/978-3-030-22605-3_8

Introduction: The ABM *Mindset*

One of the consequences of the current economic crisis was the birth of a great debate on the predictive ability of the mathematical models used in economics, particularly in the financial field (see *Economists' Grail: A Post-Crash Model*, Wall Street Journal, Tuesday, November 30, 2010). One problem is certainly the teaching: how to teach the dynamics of new (and potentially dangerous) financial instruments?

In this work (i), we show some limitations of classical economic learning, which uses a description of the dynamics through mathematical models; (ii) we propose an alternative approach, organized around an agent-based simulation, to integrate the mathematical approach with this "computational" framework. In our model, an operator observes, by means of a simulation (CDS Manager), the behavior of other agents (artificial) in its own market and is able to control since the behavior of the latter "dirty" the significance of the variables on which is taking its decisions, for example, price. CDS Manager can play two different roles: the first is a typical educational role (Padovan et al. 2002; Parisi et al. 2003). The human operator looks if his choices decrease its ability to correctly "read" the dynamics. The second role is informative: you can let the simulator use only artificial agents, generating graphical reports about what is happening and what are the predictions that agents are formulating. Furthermore, we can show some results obtained by CDS Manager on small classes of PhD students in economic/social field.

Our framework is *agent-based simulation*. What do we mean by that? In agent-based modeling (ABM), a system is modeled as a collection of autonomous decision-making entities called agents. Each agent individually assesses its situation and makes decisions on the basis of a set of rules (Cecconi 2007, 2008; Cecconi et al. 2007, 2010; Cecconi and Parisi 2007). Agents may execute various behaviors appropriate for the system they represent—for example, producing, consuming, or selling. Repetitive competitive interactions between agents are a feature of agent-based modeling, which relies on the power of computers to explore dynamics out of the reach of pure mathematical methods. At the simplest level, an agent-based model consists of a system of agents and the relationships between them. Even a simple agent-based model can exhibit complex behavior patterns and provide valuable information about the dynamics of the real-world system that it emulates. In addition, agents may be capable of evolving, allowing unanticipated behaviors to emerge. Sophisticated ABM sometimes incorporates neural networks, evolutionary algorithms, or other learning techniques to allow realistic learning and adaptation. ABM is a *mindset* more than a technology. The ABM mindset consists of describing a system from the perspective of its constituent units. A number of researchers think that the alternative to ABM is traditional differential equation modeling; this is wrong, as a set of differential equations, each describing the dynamics of one of the system's constituent units, is an agent-based model. A synonym of ABM would be microscopic modeling, and an alternative would be macroscopic modeling (Davis et al. 2003, Deng and Tsacle 2003).

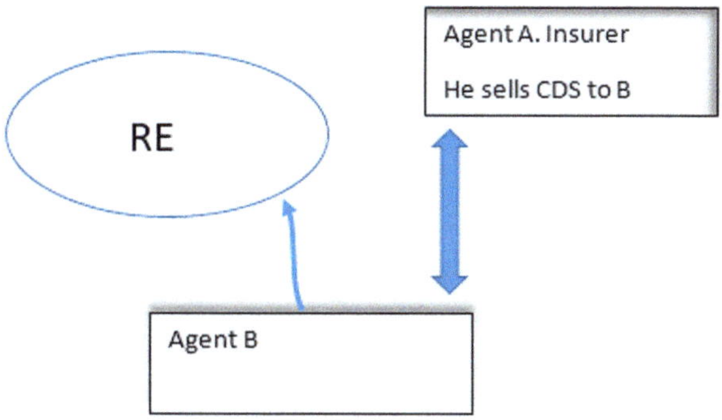

Fig. 8.1 The curved arrow indicates that B has made a loan (say $ 100) to RE, which lasted for 5 years. At the end of the first year, for some reason, B is concerned that RE may not pay back the debt and makes sure to buy a CDS from A. The solid arrow indicates the fact that you have bought insurance (CDS). B must pay an annual premium to A, say 5% of the loan amount (5 euro). Now if you were to do RE default, A is committed to repay the whole amount to B (100 euro)

The CDS Puzzle

CDS means credit default swap, and it is defined as a contract between two operators regarding the ability or otherwise of a third party, called the reference entity (abbreviated RE), to repay a debt. Clarify with an example, referring to Fig. 8.1.

Early in the process, things take place exactly as in a normal contract of insurance: paying 5 per year, B is certain that, in the event that RE decides not to honor its commitments, the full amount (100 euro) will come back. *Status quo* is complicated by the appearance of C (Fig. 8.2). For some reason, C decides to buy the CDS from B and offers a higher premium than it is to pay B to A, say 7%. If B accepts the offer, the insurance passes to C, and B starts earning 2% (7% from C, 5% to A) annually. At this point, the situation can move forward. C could sell the CDS to A, thus closing the circle. Or C sells the CDS to other agents, creating a potentially very complex network topology.

The most common econometric models describe the dynamics of the premiums (and therefore the cost of CDS) based on various assignments of probability of default. A complete description is on *Salih N. Neftçi*, *Principles of Financial Engineering*, Academic Press, 2008. The problems of the classical approach to this type of problems are of two types: the first is linked to the difficulty of explicitly describe the network of dependencies. The network is a critical factor because the presence of open loops generates the possibility that, in the event of the failure of some insurers, the amount of refund is much higher than the sum insured. The second aspect is the behavior of operators: for example, mathematical models tend to ignore that "... some reason ..." pushing traders to speculate on CDS are related to some kind of feedback. The simple news that there is a *desire* to purchase a CDS

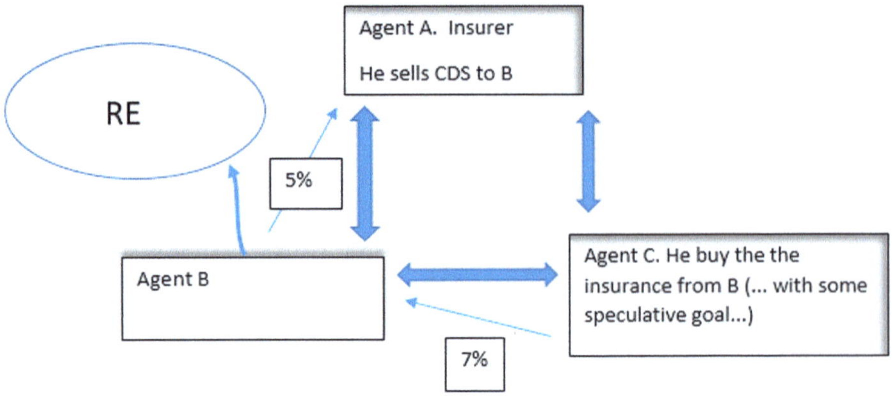

Fig. 8.2 The CDS network is growing. Agent B sells his protection to C, which agrees to pay a premium of 7%. This allows B to obtain a gain of 2%, but giving up the insurance on the debt. To be considered: C has purchased the CDS, but does *NOT* have any debt; its position is a so-called naked; he can only hope on RE default (... or to sell the CDS at a higher price…). And what happens as the network grows? What systemic risks we encounter? We must remember that A, B, C, etc. can potentially default once that is required to implement the insurance

leads to negative evaluation on the quality of debt, and indirectly to the rise in the price of CDS, and so on.

CDS Network

The controversial aspect about a CDS that makes the analogy with an insurance contract of limited use is that the buyer of a CDS need not own any underlying security or have any credit exposure to the reference entity that needs to be hedged. The so-called naked CDS buy position is, therefore, a speculative one undertaken for pecuniary gain from either the cash settlement in the event of a default or a chance to offset the CDS purchase with a sale at an improved CDS spread. This implies that gross CDS notional values can be several (5–10) multiples of the underlying value of the debt obligations of the reference entity. It has been widely noted that naked CDS buyers with no insurable interest will gain considerably from the bankruptcy of the reference entity. Agent C has an incentive to short the stock of the reference entity to trigger its insolvency in order to collect the insurance to be paid up on the CDS. A naked CDS buy position is equivalent to shorting the reference bonds without the problems of a short squeeze that raises the recovery value of the bonds (and lowers the payoff on the CDS) when short sellers of the bonds have to "buy back" at time of the credit event.

Other Two Derivative Instruments: ABS and CDOs

"Asset-backed security" (ABS). An asset-backed security is a security whose value and income payments are derived from and collateralized (or "backed") by a specified pool of underlying assets. The pool of assets is typically a group of small and illiquid assets that are unable to be sold individually. Pooling the assets into financial instruments allows them to be sold to general investors, a process called securitization, and allows the risk of investing in the underlying assets to be diversified because each security will represent a fraction of the total value of the diverse pool of underlying assets. The pools of underlying assets can include common payments from credit cards, auto loans, and mortgage loans to esoteric cash flows from aircraft leases, royalty payments, and movie revenues and *CDS*.

"CDOs." Collateralized debt obligations (CDOs) are a type of structured asset-backed security (ABS) with multiple "tranches" that are issued by special purpose entities and collateralized by debt obligations including bonds, loans and CDS. Each tranche offers a varying degree of risk and return so as to meet investor demand. CDOs' value and payments are derived from a portfolio of fixed-income underlying assets [citation needed]. CDO securities are split into different risk classes, or tranches, whereby "senior" tranches are considered the safest securities. Interest and principal payments are made in order of seniority, so that junior tranches offer higher coupon payments (and interest rates) or lower prices to compensate for additional default risk.

CDS Manager: The General Framework

CDS Manager shows the evolution of risk in a network of sellers of bonds and insurance on the debt (CDS). It can also be used with three different *modalities*:

1. "Training." We could show the growth of a network of economic agents who buy/sell bonds and CDS. As the network grows, we shows the topological parameters and calculated an index that measures what would happen in case of default of one or more agents (Kim and Baylor 2006).
2. "The market" A human agent can buy/sell insurance trying to gain money by the spread. In this modality, the other agents run *economic* behaviors (e.g., buying bonds).
3. "Challenge." Finally there is the chance to play against the computer in an attempt to create an insurance network that is capable to resist a certain degree of structural failures.

Figure 8.3 shows the main interface.

1. It is the area where students "create" the market, specifying the number of operators and their characteristics.

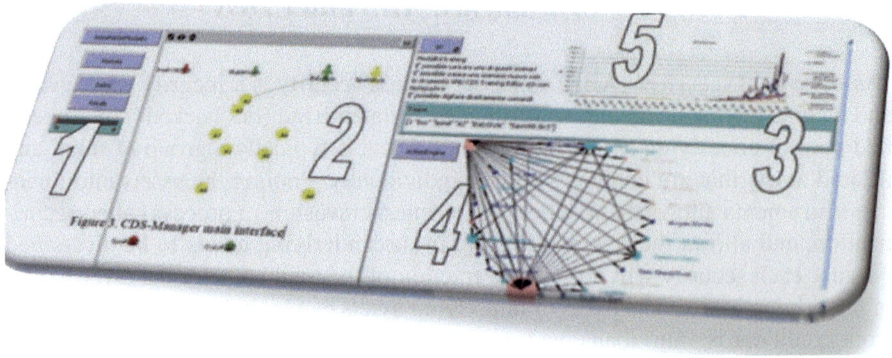

Fig. 8.3 CDS Manager main interface

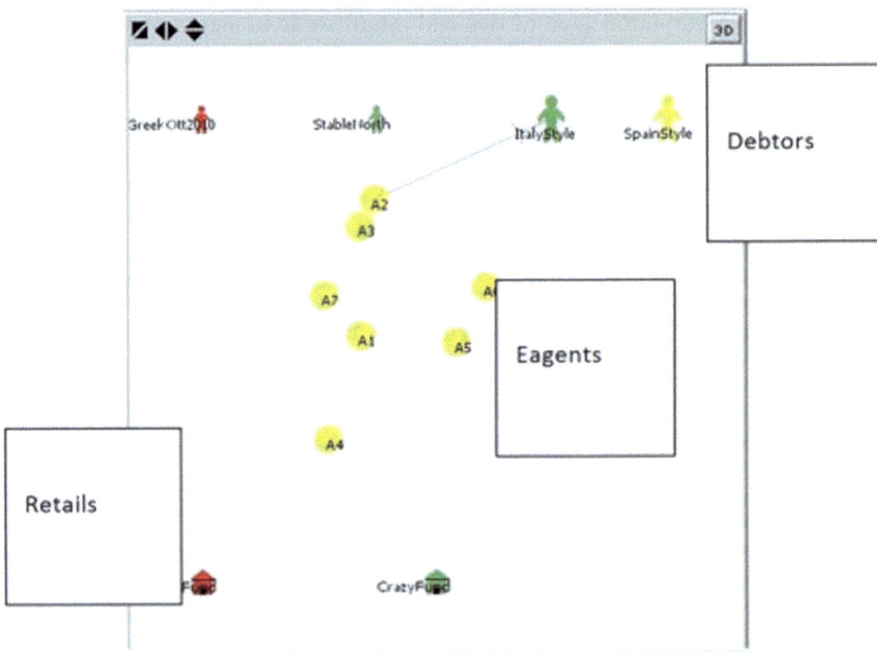

Fig. 8.4 The CDS Manager agents

2. The market (see Fig. 8.4). There are three types of agents: the *Debtors*, i.e., agents capable of opening debt position (issuing bonds); *Eagents*, that is, agents who can (i) buy bonds, (ii) issue structured bonds, and (iii) sell/purchase insurance on debt (CDS); and finally *Retails*, i.e., agents who can purchase bonds and issue structured bonds, BUT NOT BUY insurance.

The main property of *Eagents* is money (i.e., capitalization). The dynamic of money is the trigger that can set the property Default? (Default? = True means that the *Eagents* is in default state; it is no longer able to guarantee some cash flow). The properties of *Debtors* are list of bonds with duration (in years), price (default is pair), interest rate, and a valuation of default risk. In addition, *Retails* has the property RiskAttitude, a valuation about the risk attitude of the hedge fund (e.g., if RA is high, the edge fund could open a CDO position with a lot of "senior" tranches).

3. The CDS Manager engine. This is the core of the simulator. We wrote a script language to describe the behavior of the agents, called CDSBLanguage (CDS behavior language) 3. By CDSB, you can describe the behavior of the agents (during the time); for example in Fig. 8.3, we see instruction 1: an order to *Eagents A2* to buy a bond from the *Debtors ItalyStyle* (5 years, high risk). A CDSB script is called *scenario*. The students can create new scenarios or load scenarios from a library. The button GO runs the scenarios, sets the capitalizations, and modifies the behavior of artificial agents.
4. The students can load a real network to create an implementation of real economic networks.
5. The students can see the evolution of risk parameters (exposition, average of capitalizations, diameter of the network, and so on).

CDS Manager in Action

We finished the implementation of CDS Manager engine in early 2011, and since then we normally used three types of courses: Programming NetLogo, courses on the structure of economic networks, and courses in computational economics.

This is just the beginning of a real trial for CDS Manager in class students in economics or, more generally, of economic operators in the field of financial derivatives based on risk. Initial results are encouraging; however, during a mini-master of computational economics held in Rome in September of 2011, we asked to comment on one of the main stream articles on the issue, CDS, *Too Interconnected To Fail: Financial Contagion and Systemic Risk In Network Model of CDS*, Sheri Markose1, Simone Giansante∗, Mateusz Gatkowski ∗ and Ali Rais Shaghaghi∗, ∗Centre for Computational Finance and Economic Agents (CCFEA), 1Economics Department, University of Essex, UK.

Our assessment was that it was a complex article, on various levels, and that his understanding had to go through a very thorough reading. The comments were aimed at responding to a series of open questions that attempted to capture the understanding of some key concepts of the article: (a) the presence of open networks increases systemic risk, (b) the presence of operators selling structured products based on CDS increases systemic risk, and (c) the *bear run*, or the attempt to buy a CDS in the hope of increasing their value, can trigger a reduction in quality of debt insured4. The group of students (22) has dealt with this article during a morning (it was a surprise, a substitution in the normal course of lectures), after using for 3 days CDS Manager with scenarios very similar to those described in the article. The students reported comments quite in line with the topics identified. It seems to us an encouraging result, but we're gearing up for text-mining tools to *validate* this test.

Conclusions

This paper describes the implementation of an agent-based educational framework, CDS Manager. CDS Manager provides an interactive learning environment that allows students to (1) construct their markets, (2) reflect upon the quality of their understanding, and (3) test it out in the dynamically generated simulation games. The CDS Manager framework stresses reflection as the critical component of thinking processes. With an editable rule-based reasoning engine, CDS Manager makes it easier for developers to model domain knowledge and develop simulation games that are otherwise time-consuming (Edmondson 2002).

Currently CDS Manager is in a pre-engineered version, with most of the procedures written in NetLogo. We are working (and looking) for resources in order to close a beta version with extensions written in Java.

References

Cecconi, F. (2007). Financial markets and norms. *ISTC CNR, conf.*

Cecconi, F. (2008). Agent based simulation for social studies, Technical report, *10th European Agent Systems Summer School*. 5–9th of May, Lisbon, Portugal., Tech.

Cecconi, F., & Parisi, D. (2007). Asymmetric pricing: An agent based model. In *Proceedings of the IASTED International Conference on Modelling and Simulation (MS 2007)*, February 13–16, Montreal, CANADA, pp. 380–385.

Cecconi, F., Caligiore, D., Baldassarre, G., & Parisi, D. (2007). Interacting networks as model of cultural change. In *Proceedings of Computational Methods for Modeling and learning in Social and Human Sciences (MASHS 07)*, May 10–11, Brest, France, pp. 190–199.

Cecconi, F., Campenní, M., Andrighetto, G., & Conte, R. (i. p. 2010). *What do agent-based and equation-based modelling tell us about social conventions: the clash between ABM and EBM in a Congestion Game Framework*, JASSS 1, on-line.

Davis, J. S., Hecht, G., & Perkins, J. D. (2003). Social behaviors, enforcement, and tax compliance dynamics. *The Accounting Review, 78*(1), 39–69.

Deng, P. S., & Tsacle, E. G. (2003). A market-based computational approach to collaborative organizational learning. *The Journal of the Operational Research Society, 54*(9), 924–935.

Edmondson, A. C. (2002). The local and variegated nature of learning in organizations: A group-level perspective. *Organization Science, 13*(2), 128–146.

Kim, Y., & Baylor, A. L. (2006). A social-cognitive framework for pedagogical agents as learning companions. *Educational Technology Research and Development, 54*(6), 569–596.

Liu, J. I. C., Yun, D. Y. Y., & Klein, G. (1990). An agent for intelligent model management. *Journal of Management Information Systems, 7*(1), 101–122.

Padovan, B., Sackmann, S., Eymann, T., & Pippow, I. (2002). A prototype for an agent-based secure electronic marketplace including reputation-tracking mechanisms. *International Journal of Electronic Commerce, 6*(4), 93–113.

Parisi, D., Cecconi, F., & Natale, F. (2003). Cultural change in spatial environments: The role of cultural assimilation and internal changes in cultures. *The Journal of Conflict Resolution, 47*(2), 163–179.

Radicchi, F., Castellano, C., Cecconi, F., Loreto, V., Parisi, D., & Parisi, G. (2004). Defining and identifying communities in networks. *Proceedings of the National Academy of Sciences of the United States of America, 101*(9), 2658–2663.

Yang, H.-D., Mason, R. M., & Chaudhury, A. (2001). The internet, value chain visibility, and learning. *International Journal of Electronic Commerce, 6*(1), 101–120.

Chapter 9
A Decision-Making Model for Critical Infrastructures in Conditions of *Deep Uncertainty*

Juliana Bernhofer, Carlo Giupponi, and Vahid Mojtahed

Abstract In this work, we develop a set of tools able to analyse different options in a generic case of critical infrastructure development with the consideration of climate change adaptation needs. As an example that will guide us through the process, we refer to a decision problem related to the hydraulic regulation of a strategic infrastructure (e.g. an airport, a power plant, a train station or a logistic centre) exposed to uncertain future climatic extremes.

Keywords Robust decision-making · Decision-making model · Critical infrastructures · Deep uncertainty · Climate change

Introduction

Climate and socio-economic changes are amongst the most important drivers affecting the evolution of social and ecological systems (SES), but the possibilities of predicting their future trajectories are very limited. The high level of uncertainty about the future evolution of drivers and systems makes the assessment of risk and resilience a challenging task for analysts and decision-makers. The uncertainty associated to those phenomena has been defined as *deep*, because knowledge is limited both in

J. Bernhofer (✉)
Free University of Bozen-Bolzano, Bolzano, BZ, Italy
e-mail: juliana.bernhofer@unibz.it

C. Giupponi
Department of Economics, Ca' Foscari University of Venice, Venezia, VE, Italy
e-mail: cgiupponi@unive.it

V. Mojtahed
Venice International Centre for Climate Studies, Ca'foscari University of Venice, 30123 VE, Italy

Avanade UK Ltd., London, EC4M 6XH, UK
e-mail: vahid.mojtahed@avanade.com

© Springer Nature Switzerland AG 2019

F. Cecconi, M. Campennì (eds.), *Information and Communication Technologies (ICT) in Economic Modeling*, Computational Social Sciences,
https://doi.org/10.1007/978-3-030-22605-3_9

statistical and epistemological terms, in that not only the quantification of these uncertainties is difficult (e.g. unknown distributions) but also the main concepts such as resilience are not well defined and universally accepted. Moreover, there is an inherent uncertainty attached to the choice of the best model for the measurement of resilience in a non-informed context and to the definition of a satisficing level of performance in a complex system. As stated above, in the existing literature, such premises are generally referred to as a situation of *deep or severe uncertainty* (Lempert and Collins 2007; Lempert and Kalra 2013). Therefore, the need emerges to assist decision-makers not by providing them with an optimal solution based on known past trends or a few plausible future scenarios but rather with an analysis able to provide ranges of possible future outcomes under wide sets of plausible scenarios in order to identify more robust solutions. As compared to optimal ones, robust solutions are those that show relatively limited cases of failure, as defined and agreed by stakeholder in advance, in a high number of possible diverse, yet feasible, conditions.

A decision-making approach based on the identification of robust solutions (Robust Decision-Making or RDM) has been implemented in several different contexts, such as agriculture, resource management and strategic infrastructure. In this work, we take the latter as an example. Public and private investment in maintenance, protection and in the development of strategic, long-lasting infrastructure, such as networks of transportation, water, communication and energy, are crucial to the functioning and prospering of modern societies. During recent years, particular attention has been paid to *critical infrastructures*, i.e. assets that are essential to the functioning of current and future societies as a whole, such as electric power plants, telecommunication and transportation networks, which are considered "too important to fail".

Impacts of extreme weather events, like the heavy flooding of New York's LaGuardia Airport as a consequence of Hurricane Sandy have caused severe social and economic damage due to the disruption in commercial and transportation services. This and similar events have brought the attention of decision-makers to assessing and enhancing the resilience of critical infrastructures. Within this context, the concept of resilience, in short, can be defined as the ability of a system to withstand shocks by exogenous drivers, or to recover in a timely manner, thus returning to its original or improved state of functionality. Critical infrastructures are usually characterized by considerable initial and maintenance costs and must therefore be carefully chosen by considering potential benefits, trade-offs and interactions with the existing and future conditions.

In this work, we develop a set of tools able to analyse different options in a generic case of critical infrastructure development with the consideration of climate change adaptation needs. As an example that will guide us through the process, we refer to a decision problem related to the hydraulic regulation of a strategic infrastructure (e.g. an airport, a power plant, a train station or a logistic centre) exposed to uncertain future climatic extremes. The proposed approach was originally designed for assessing the robustness of project alternatives for the defence from flood risks due to extreme rainfall events, but it would be suitable to consider also other sources of uncertainties in infrastructure design, including other climatic, as well as financial and political stressors.

First, we employ an existing multi-criteria analysis (MCA henceforth) tool, i.e. the mDSS software (Giupponi et al. 2004), with which we evaluate a catalogue of decision-relevant and potentially sensitive indicators (e.g. investment costs, resilience enhancement, environmental impact). We choose this approach because of its flexibility in dealing with a multitude of factors related to the infrastructure and due to lack of formal mechanical/mathematical models for assessing resilience. The indicator list is usually set up by consulting a board of multidisciplinary experts involved in the particular decision problem, who also assign relative importance weights to each element. The input analysis matrix is obtained by assigning a score to each menu point in the catalogue under the proposed infrastructural solutions. Such infrastructural solutions in the case of a protection plan against extreme weather events could include elevation of the infrastructure, increase of the drainage capacity or relocation. The MCA is initially run by using criterion scores averaged across all expert responses that can have different views towards realization and measurement of uncertainties and its impacts. With this first benchmark result at hand, we proceed with a multi-scenario sensitivity analysis by introducing various sources of uncertainty, which are expert criterion scores and importance weights, risk attitude of the decision-maker and quantitative criteria (e.g. costs).

We then employ our computational tool developed in MATLAB that combines numerous (hundreds, thousands or hundreds of thousands depending on the context and magnitude of uncertainty) uncertain multi-criteria decision analysis matrices to approximate the performance of each alternative and create a range of possible outlooks. The deficiencies of each plan under different sets of scenarios are determined and, based on the above assessment, the robustness of each solution is defined, following the methodology of Rosenhead (1980a, b). The most robust solution is the solution that is considered as preferable in most of the simulated alternative feasible scenarios. Finally, as a sensitivity analysis that also provides intuitive visual support, we apply a CART (Classification And Regression Trees) algorithm, which allows for an identification of patterns of critical score values able to overturn the final ranking of decision alternatives. The combination of these components results in a comprehensive and intuitive decision support system that helps the decision-maker to mitigate the impacts of uncertainty and to increase the system's resilience by identifying the most robust plan, which is to be interpreted as the option that ranks best under most of the simulated alternative scenarios.

In what follows, sect. 2 describes the methodology of multi-criteria decision-making and of the simulation tools, in sect. 3 we conduct a step-by-step sample analysis and section 4 concludes.

Methodology

In a recent work on climate change adaptation (Giupponi 2014), possible decision-making situations were categorized according to time scales in the context of water resource management (WRM henceforth), ranging from real-time problems, such as weather alerts, to long-term ones, which also consider the possible evolution of

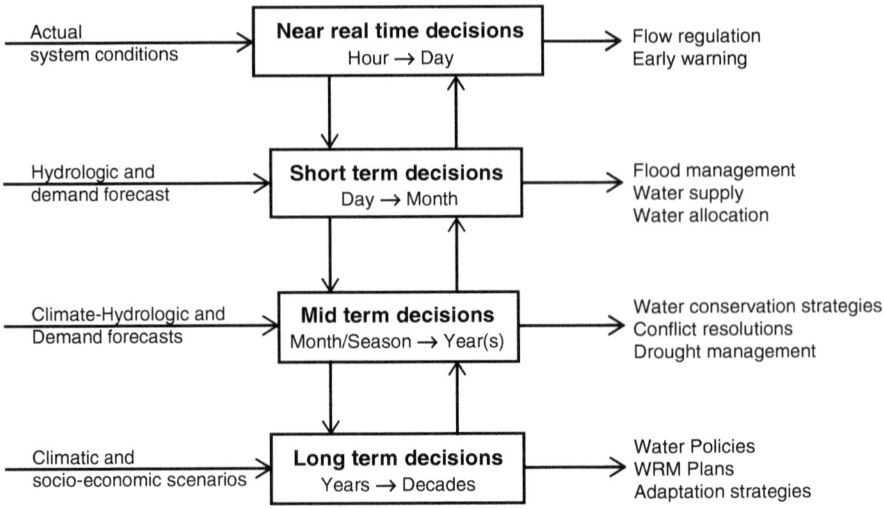

Fig. 9.1 Categorization of decision levels by time frames for water resource management. (WRM) (From Giupponi 2014 adapted from Georgakakos 2007)

socio-economic and environmental variables (see Fig. 9.1). The main concern of this work is that of long-term decisions and, in particular, the context of adaptation strategies. Scenario analysis plays a fundamental role in exploring the expected performance for the possible alternative solutions or options when affected by different sources of uncertainty.

The decision in such cases typically starts from the identification of feasible solutions considering multiple future scenarios. The decision-making process is aimed at identifying one or more preferable alternatives by considering the following elements:

- A set of *alternative options*, which are the different solutions to the decision problem under analysis
- A catalogue of *indicators* for the evaluation of each of the alternative options
- A range of sources of *information and data*, from modelling outcomes to qualitative information provided by experts and stakeholders
- A comprehensive definition of the *sources of uncertainty*
- *Constraints* or acceptable lower or upper bounds of the criteria defined earlier (e.g. budget constraints on the costs, constraints in terms of space, legal constraints)
- *Objective functions* which define whether a certain criterion or indicator should be minimized (e.g. costs, environmental impacts) or maximized (e.g. reliability, flexibility)
- A set of *plausible scenarios* that represent the counterfactual environments in which each the performance of each option is going to be assessed
- A *preference structure* that defines the relative importance of each criteria in contributing to the final score

Decisions for the Adaptation to Global Changes: The NetSyMoD Method

As early as the 1970s, interactive computer systems have been proposed to help decision-makers in the use of data and models when solving complex and unstructured problems. In particular, Gorry and Scott-Morton (1971) and Keen and Scott-Morton (1978) proposed to couple intellectual resources of individuals with computer skills to improve the quality of decisions.

Recent guidelines of the European Commission underline the importance of "climate-proofing" actions at EU level which consist in the promotion of adaptation in vulnerable sectors, such as agriculture and fishery, and in making Europe's infrastructure more resilient to climate change (European Commission 2014). Computer-assisted Decision Support System (DSS henceforth) represent an invaluable tool when it comes to investment decisions for strategic infrastructure which are characterized by significant financial, environmental, social and political responsibilities and must last for future generations. The advantage of supporting the decision process with dedicated software lies in the replicability, visualization and traceability of the results. Only when it is possible to conduct counterfactual analyses which can be compared, the best solution in terms of robustness can be chosen. Also, simulation results can be continuously updated and refined as new or more precise data becomes available over time.

A realistic DSS, however, must consider the human component an integral part, as decision-makers may have different time preferences or risk attitudes. Also, strategic investment decisions are frequently based on highly specialised topics wherefore the technical knowledge and option evaluation is most efficiently provided by experts in the field. Finally, the computer-assisted DSS should also allow for the input of external specialised studies (e.g. hydraulic models, climate assessments) and bottom-up sensitivity analyses able to identify critical areas and scores that could revert outcomes.

For our sample study we present the NetSyMod (Network Analysis-Creative System Modelling-Decision Support) (Giupponi et al. 2008)[1] methodology which has proven useful in numerous decision problems linked to WRM and CCA. The resulting software mDSS (Giupponi 2007) is in line with the European guidelines and was developed with the aim of comparing alternative options through multi-criterial analysis on one hand while featuring a sufficient level of flexibility for the integration of the human component as well as of external studies on the other. The overall process structure of the NetSyMod method is summarized in Fig. 9.2. After the activation phase, a significant part is dedicated to the exploration of the problem with the involvement of stakeholders and experts. Only in step 5 the actual quantitative analysis of the data takes place in order to choose the preferable option.

[1] See also http://www.netsymod.eu/.

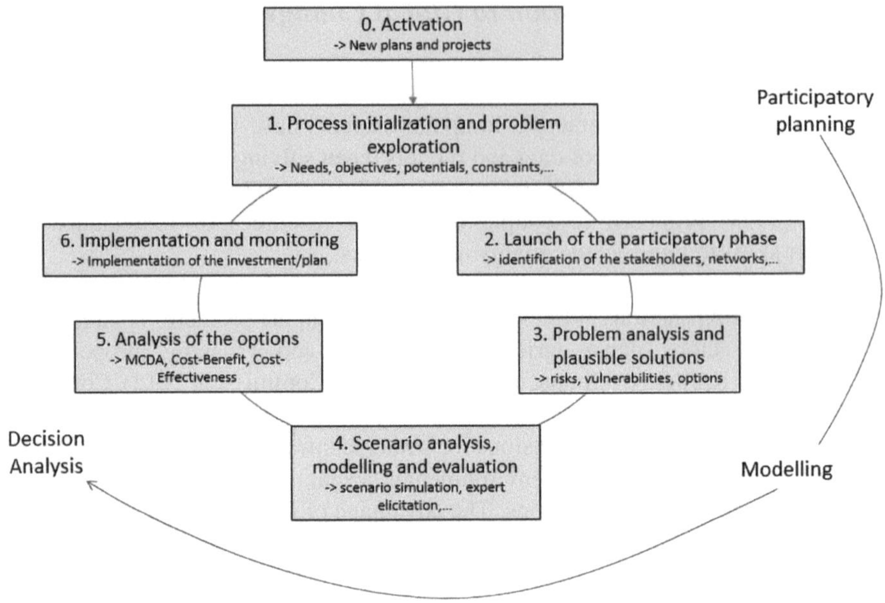

Fig. 9.2 Steps of the NetSyMod procedure. (Adapted from Giupponi 2014)

Multi-criteria Decision Analysis (MCDA)

A decision-making problem in the context of investment-related choices, be it of public or private interest, has two main questions to be addressed: first, whether the intervention is financially convenient, and second, whether the proposed solution is the best among all possible alternatives. Therefore, optimal choices are the result of comparing alternatives and considering trade-offs.

The fundamental driver that led to the development of MCDA approaches was the recognition that investment choices are seldom made by considering only a single criterion but are always multidimensional. Moreover, the multidimensionality of the choice does not characterize only the initial phases of the evaluative process but permeates it completely. The process of choice must consist in the analysis of several options until the preferable one emerges. A possible representation of the multi-criteria decision problem is shown in the graph of Fig. 9.3.

Axes a_1 and a_2 represent the principal aims pursued with the investment, whereas the alternatives are represented by letters A to F. We can now eliminate those alternatives which, in the presence of a rational decision-maker, would never be considered as they are dominated by better solutions. In this spirit we see from Fig. 9.3, that investment options B, F and D can be discarded.

$$\text{C vs B:} \quad a_{1C} > a_{1B} \quad \textit{and} \quad a_{2C} = a_{2B}$$
$$\text{C vs F:} \quad a_{1C} > a_{1F} \quad \textit{and} \quad a_{2C} > a_{2F}$$
$$\text{E vs D:} \quad a_{1E} > a_{1D} \quad \textit{and} \quad a_{2E} = a_{2D}$$

Fig. 9.3 Representation of a decision problem with utility function defined by two criteria (a_1 and a_2)

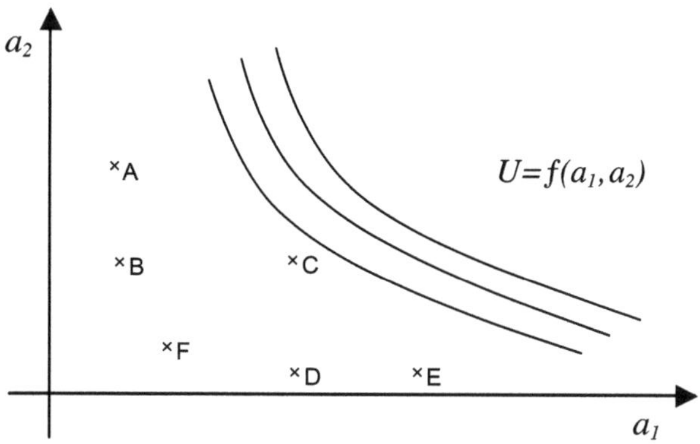

In other words, C Pareto dominates[2] B and F, while E dominates D.

However, the examination of the graph also shows that the principle of dominance used to discard investments B, F and D proves not useful in determining an optimal ranking of the remaining alternatives A, C and E. As a result, to continue the analysis, it is necessary to formulate hypotheses about the structure of the decision-makers' preferences. Assuming that these preferences are represented by the convex and monotonic utility function $U = f(a_1,a_2)$, we can now determine the preferable solution with respect to U. Along each indifference curve of the set U, the alternatives have the same total value to the decision-maker, as, while moving along the curve, a_1 and a_2 are trading off perfectly between each other. If we assume that increasing values of a_i are desirable, then alternatives that lie on higher indifference curves are preferable. Hence as $U_C > U_E > U_A$, C becomes the preferable alternative (Fig. 9.4).

The main question that arises in the evaluation of investments by the multi-criteria method is therefore the development of appropriate procedures to represent the preferences of the decision-makers to be applied to a set of alternative investments that are not Pareto dominated. More specifically, in the present case, the decision-making problem can be defined as a choice between n alternatives A_i by assessing a finite number k of criteria a_j where each alternative has a certain performance score a_{ij}.

In this work, we went beyond the usual additive approach, where the aggregated indices are the result of summing normalized and weighted values. Although this method (defined as "weighted sum" or *Simple Additive Weighting*, SAW, henceforth) is simple and straightforward to understand, it is completely compensatory: the result of aggregating an excellent and a bad value is the same as when two average values are aggregated (this situation is usually defined as ORness). At the other extreme is the case of multiplication aggregation (ANDness) which represents the

[2] The principle of Pareto dominance (Pareto optimality) is extremely useful in investment decisions and even more in the field of multi-criteria analysis as it allows to isolate those alternatives of choice with contrasting rankings in their criteria and thus identifies real conflicts (trade-offs) in the decision-making process.

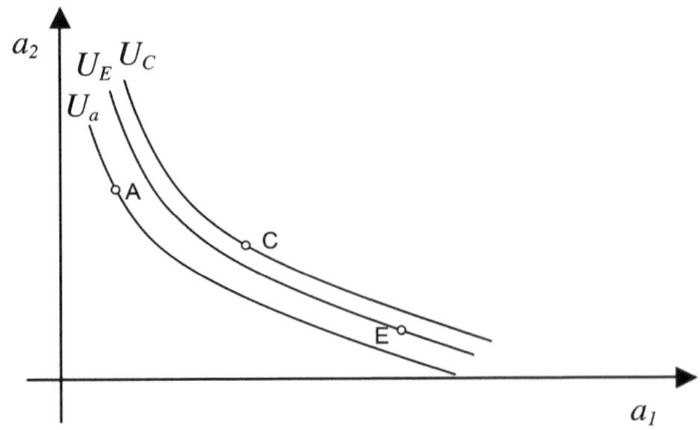

Fig. 9.4 The solution to the multi-criteria decision problem

opposite problem: when only one of the indicators has zero value, the entire aggregated index is zero. In order to overcome these limitations, the method that has been adopted is that of Ordered Weighted Averages (OWA henceforth). The OWA method (Ronaldo R. Yager 1988) aggregates the scores in a second weighting round in which new weights are applied to the ordered sequence of values previously weighted as in SAW. Note that in this case, the weights are attributed to the sequence of weighted criteria and not to each single criterion. For example, if three indicators are to be aggregated, first, their values are weighted as usual (weighted scores) and then they are sorted (sorted scores) and re-weighted with a new weight vector. This second phase allows to overcome the compensation effects of SAW and to implement the preferred degree of ANDness, with two extremes: the case of a pessimistic decision-maker in which the highest importance is given to the criterion with the lowest weighted value and the optimistic case where only the highest values count for the final score. The case in which all ordered weights are equal reproduces the SAW case. Variations in the level of asymmetry in ordered weights produce solutions that represent different levels of risk aversion. Therefore, the balance between ANDness and ORness allows for the exploration of different risk attitudes of the decision-maker.

Uncertainty is a crucial issue here, which pervades all aspects of the decision-making processes. Multiple scenarios can be considered with the construction of multiple matrices, where each matrix shows how the indicators vary—in expected terms—in different states. Climatic projections are inherently limited in the ability to simulate extremely complex phenomena, especially when they are analysed on a local scale, but also the expected trend of socio-economic variables is characterized by extreme variability in the long term.

The concept of uncertainty has different definitions in literature, and its idea is not often distinct from the concept of risk, with many researchers approaching uncertainty with the tools available for risk assessment. According to Knight (1921), risk is the calculable and controllable part of the unknown and is represented by the distribution of probabilities on a finite and well-known set of events or results. The remainder is uncertainty, which is uncontrollable and incalculable. While the definition of Knight eliminates every effort to quantify the uncertainty of a complex system as a whole, our purpose is to distinguish and arrive at the sources of uncertainty in order to provide decision-makers with a tangible measure of uncertainty.

Walker et al. (2003) suggest that uncertainty can emerge in various areas of a complex system such as context, model structure, inputs, parameters and outcomes. To address the issue, first of all we need a taxonomy that helps to identify the position as well as the degree and nature of the uncertainties. The degree of uncertainty can vary from statistical uncertainty (risk) to scenario uncertainty, declared ignorance and finally total ignorance. To a certain extent, uncertainty is also a subjective issue as it depends on the perception and satisfaction of stakeholders regarding existing knowledge. Part of this ignorance can be reduced by acquiring more information. In such cases, uncertainty is known to be *epistemic*. If we are not able to reduce uncertainty due to the natural variability of a complex system, then the uncertainty is said to be *aleatory*. Assessing the frequency of floods considering future climate change is an example of aleatory uncertainty.

During the last years, a great deal of research has developed several approaches to decision-making in uncertain conditions. Examples are robust decision-making (Bryant and Lempert 2010; Groves and Lempert 2007; Herman et al. 2015; Lempert and Kalra 2013; Lempert et al. 2006), scenario planning (Mahmoud et al. 2009; Schoemaker 1991, 1993), info-gap (Ben-Haim 2004), Bayesian network (Darwiche 2009), adaptation pathways (Haasnoot et al. 2014) and many objective robust decision-making (Kasprzyk et al. 2013). Many of these approaches have been used in the area of WRM, but they failed to consider internal interdependencies of economic factors in their uncertainty analysis. Herman et al. (2015) provided a taxonomy of the robustness frameworks and identified the similarities and differences of these methodologies in the field of urban WRM. However, to our knowledge none of these studies have considered socio-economic uncertainties or the dependencies between uncertain variables that could help the internal coherence and plausibility of the scenarios.The choice between the methodologies depends on specific theoretical considerations and cases, but before going into the design of an approach able to deal with uncertainty, the general methodological framework of the decision-making process and the modalities of risk assessment must be defined.

As a final result of the multi-criteria decision analysis, one or more options emerge as preferable ones, but in addition to produce this simple information, MCDA allows for an accurate documentation of results, assumptions, trade-offs, elements of subjectivity and uncertainties of various types, communicating them transparently with graphs, tables and statistical annexes.

As for the assessment of uncertainty, the literature shows above all a particular attention to the uncertainties of natural hazards, thus producing probabilistic mappings of dangers and risks. The tragic events that society suffers as consequences of natural hazards are instead the result of very specific interactions between the different variables and factors that make up the risk, both natural and anthropic, which must be considered in risk assessment procedures. To this end, the analysis of uncertainties gives decision-makers the opportunity to explore their decision-making spaces and to seek risk reduction measures (not only limited to structural measures) that might appear to be more robust in an uncertain future.

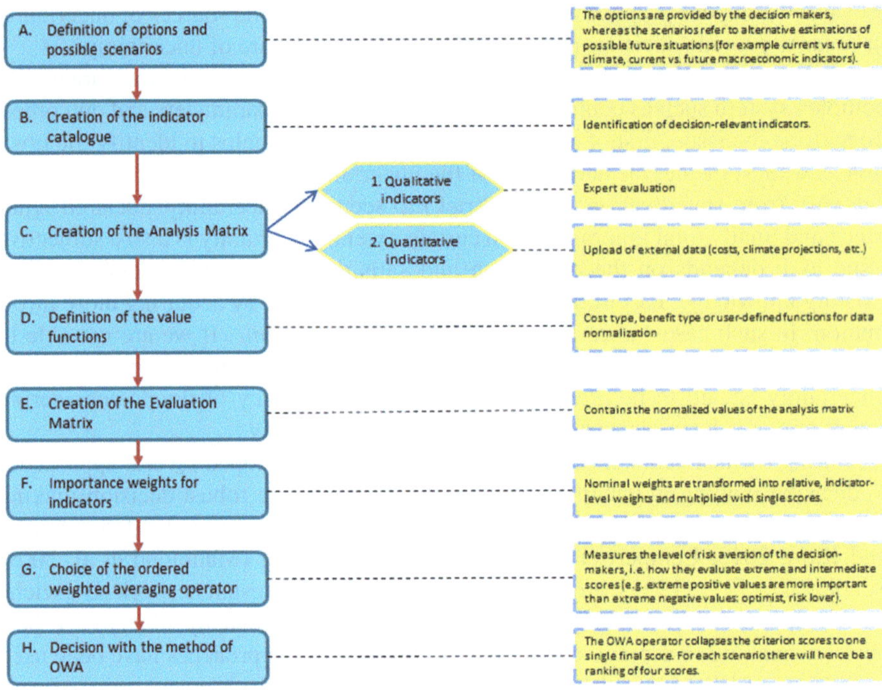

Fig. 9.5 Multi-criteria decision process with mDSS for the evaluation of alternative investment options

Implementation with mDSS[3]

Figure 9.5 shows the steps of the multi-criteria analysis which were implemented in our sample study. As a starting point of the analysis, the possible options (i.e. alternative solutions) and future scenarios are laid out. This is done together with the decision-makers and experts, who, guided by the research team, also actively contribute in determining all decision-relevant indicators.

The Analysis Matrix is the core element of the multi-criteria analysis and usually contains qualitative and quantitative indicator values for each of the possible future scenarios. Each column represents one of the alternative solutions and each row contains one indicator value. The analysis matrix has then to be normalized in order to create comparable values. The result is the Evaluation Matrix (step E) with the base values to be transformed by user-defined importance weights and OWA vectors. The OWA weights collapse the importance-weighted scores to one value for each alternative option.

[3] Download available at http://www.netsymod.eu/DSSwelcome.html.

Application to an Example of Infrastructural Investment Under Deep Uncertainty

We present a generic example of an investment decision in a strategic infrastructure subject to flood risk which may apply for example to roads, airports and telecommunication systems. The aim is to guide the reader through an applied case of decision-making under deep uncertainty and provide a viable tool to solve concrete decision-making problems. In what follows, we will describe each step of the flow chart represented in Fig. 9.6 in detail.

Activation, Problem Exploration and Information Exchange (Step 1)

The first step is a process of *mutual learning* between analysts, decision-makers, experts and other stakeholders where the problem is laid out in detail and the expectations of each party are laid out.

Moreover, the overall infrastructural system is defined, and the conditions under which this system ceases to meet its functional requirements are assessed. *Failure* (see point 1a in Fig. 9.6) is assumed to be a situation of service interruption in case of extreme weather events, which, in practical terms, translates into the maximum flood limit supported by each area. Parallel to defining under which conditions a situation of system failure would apply, also critical areas that behave differently from others are localized and highlighted. The physical model contains the network of interconnected elements and is visually supported by detailed maps.

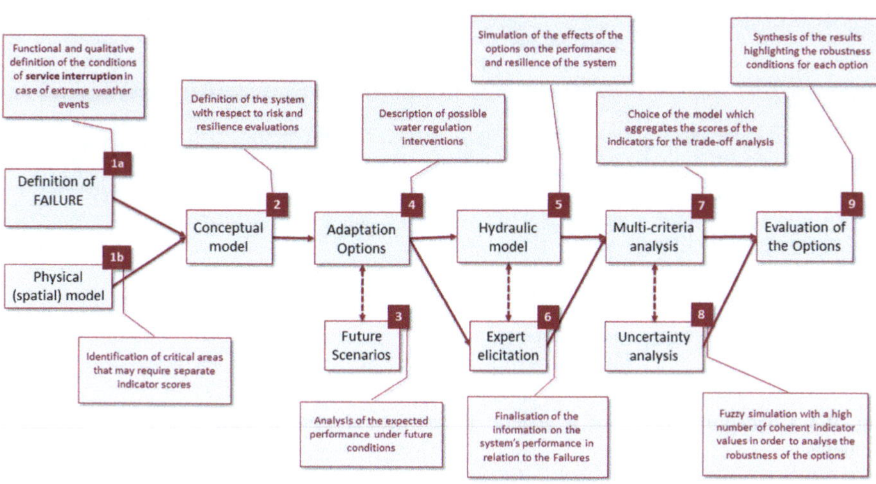

Fig. 9.6 Flowchart of the logical and temporal project sequence

Conceptual Model (Step 2)

The purpose of the conceptual model is to provide in a simple graphic design the elements, the variables and the main links that are relevant for the infrastructural functioning and which can change under the implementation of alternative WRM solutions under extreme meteorological pressure.

Figure 9.7 displays the logical links that associate risk assessment to extreme events within the infrastructural system under analysis according to the concepts of danger (hazard), vulnerability, exposure and risk and through the consideration of the adapting and coping capacity of the system alongside its resilience. The ex-ante preparedness of the infrastructural system to combat hazard and reduce its adverse impact can be defined as adaptive capacity, while the ex-post skills to cope with and overcome the impacts of the hazard considered is usually defined as coping capacity (IPCC 2012). The red boxes contain elements that are potentially affected by the investment; the blue boxes represent the different scenarios for the analysis (traffic and climate in this case).

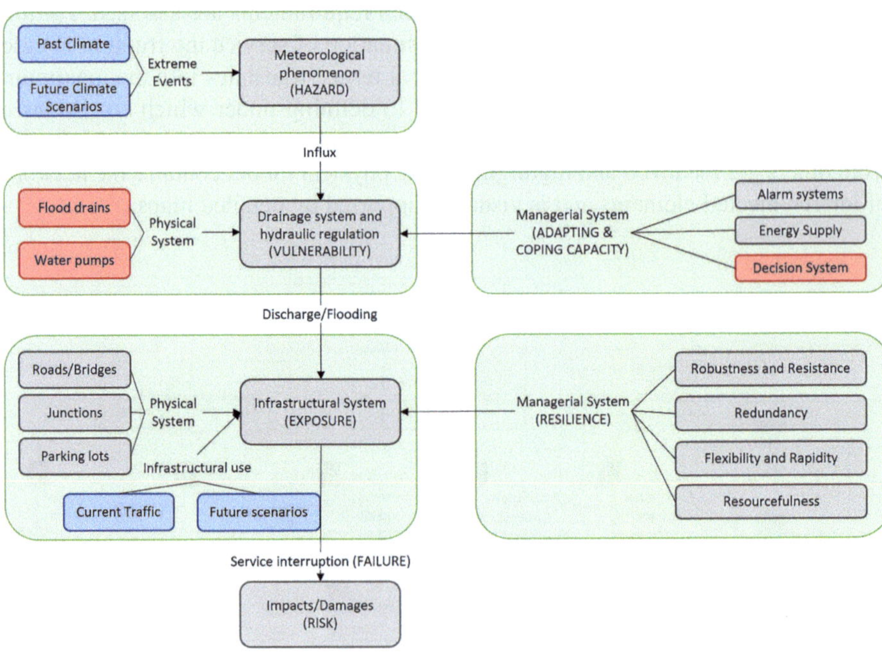

Fig. 9.7 Conceptual model of a generic infrastructural system with examples of its main elements

Future Scenarios (Step 3)

The aim of this step is to provide a plausible range of possible future "worlds". Future climatic scenarios are to be estimated by engineering professionals with climatic time series and specialized computer programs. Additional variability can be added by changing the return period of extreme events. The results are combined with the results of possible future traffic scenarios to obtain a simulation of possible overall states in a long-term perspective.

The aim of the uncertainty analysis will therefore be to analyse many combinations of climatic situations and traffic outlooks to determine whether and to what extent the different solutions considered may contribute to reducing the risk of temporary inoperability.

Adaptation Options (Step 4)

Step 4 is taken in parallel to the intertwined step 3 and defines the possible solutions to the decision problem of WRM for strategic infrastructure which, in the presented case, are the following[4]:

A1. Direct management of the water pumps and improvement of existing pumps
A2. Direct management of the water pumps and construction of an additional surge tank
A3. External management of the water pump system, construction of additional embankments and surge tanks
A4. External management of the water pump system, construction of a retarding basin

Water discharge after extreme weather events in alternatives A1 and A2 is regulated by improving existing water pumps or the construction of an additional surge tank respectively. The first two options are managed directly by the control body of the strategic infrastructure. In options 3 and 4, the idea is to outsource the water pump management and to construct additional embankments, surge tanks or retarding basins. In the hydraulic model (step 5 of Fig. 9.6), an additional strain to the drainage system was considered by hypothesizing a power outage. It becomes clearer now that the water pumps are a sensitive area with a highly variable impact based on its strategic role in the WRM process and the associated managerial responsibility.

The four options appear as a catalogue in the mDSS software (Fig. 9.8).

[4]We listed these four options for the sake of concreteness and to add realism to our case study. However, any project alternative with evident or potential trade-offs in their performances could be considered an option.

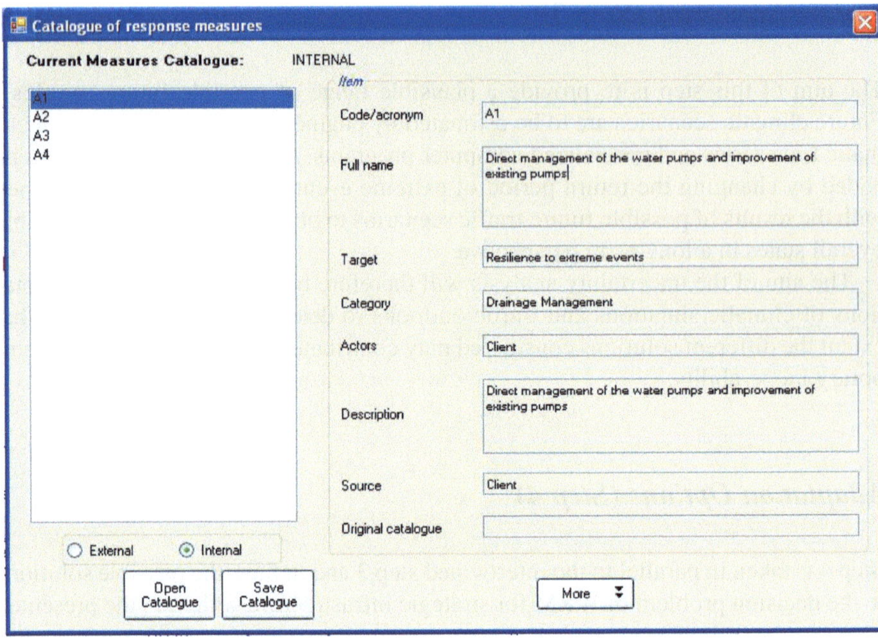

Fig. 9.8 Response option catalogue in mDSS

Hydraulic Simulation (Step 5)

A hydraulic simulation conducted by engineering professionals helps in considering all interdependencies of the system and in visualizing its behavioural pattern under different pressures in several future scenarios. The model is parametrized with time series of past weather events, traffic data, recorded data from comparable existing projects, etc. and provides a visualization of the extension and duration of flooding during different extreme weather events.

Moreover, another dimension was introduced to the scenario analysis in order to assess the consequences in terms of operability following the combination of extreme climatic phenomena and other factors of shock to the system, such as a power outage which may interrupt or limit the functionality of the water pumps.

Expert Elicitation (Step 6)

It is clear that the information from the simulations of the hydraulic model and the evaluation of management and investment costs are of paramount importance in order to choose the preferable option, but they are not sufficient to a comprehensive

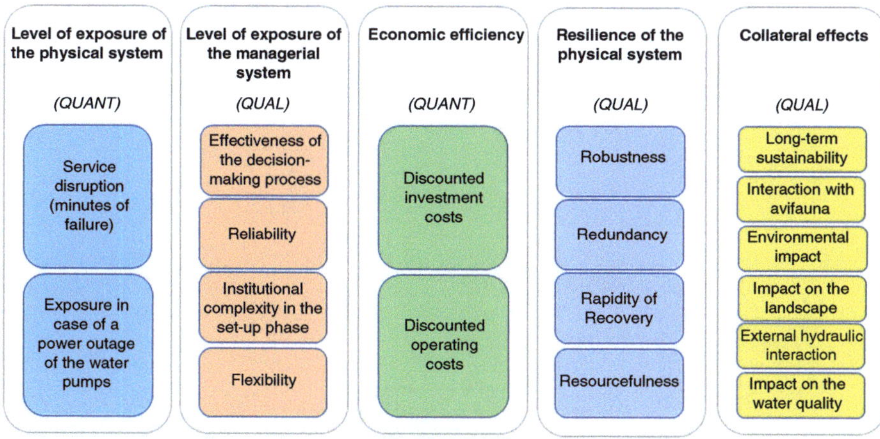

Fig. 9.9 Qualitative and quantitative indicators

understanding of the decision-making procedures. The quantitative data has to be considered alongside qualitative indicators, where the latter of which are defined through the elicitation of expert knowledge. Figure 9.9 shows the final list of the identified quantitative and qualitative macro-criteria and the indicators used for the evaluation.

Once the objectives and methodology of the data collection have been defined in detail, experts are called to assign both performance scores and weights of importance to each indicator to be used as input vectors for the classic multi-criteria analysis presented in the next section.

Multi-criteria Analysis (Step 7)

Following the procedure represented in Fig. 9.5, we can now impute the values of the analysis matrix which contains the non-normalized values of the qualitative and the quantitative indicators. Figure 9.10 represents steps C to E of the mDSS flow of Fig. 9.5: the analysis matrix is transformed into the evaluation matrix by applying the chosen value functions for normalization. Here we decided to use min-max normalization and chose the cost or benefit type for each indicator. The result is a matrix containing normalized comparable values for the current and the future climate.

Data for quantitative indicators stems from internal and/or external ad hoc studies, whereas qualitative indicator values are provided by experts by means of questionnaires. The answers were given on Likert scales from 1 to 5.[5]

[5] To add additional variability to the simulation database, Likert scales could also be fuzzy (Fourali 1997), i.e. it could be made possible to choose more than one value, creating in this way a range instead of a spot value.

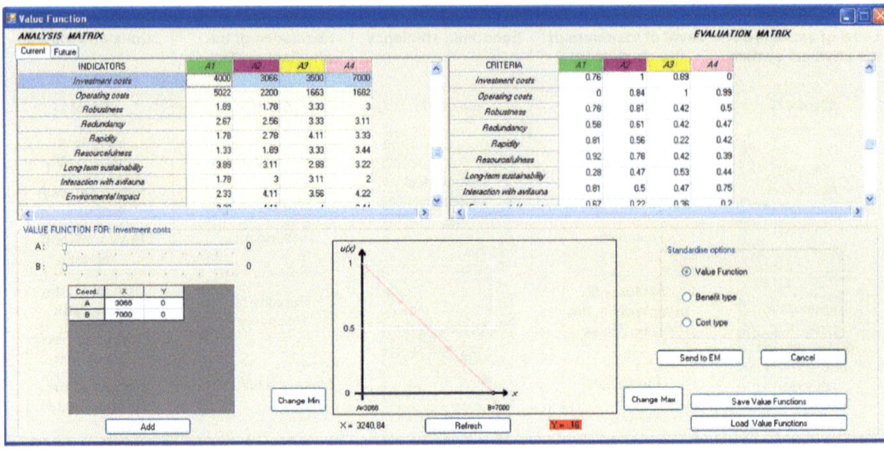

Fig. 9.10 From the analysis matrix to the evaluation matrix

Next, the importance weights assigned by experts are applied to the normalized values. Based on the composition of the expert group, there is a vast variety of possibilities how to aggregate individual importance weight vectors. Besides straightforward averaging over the whole group, different between-expert weights could be assigned based on the position of the expert in the decision-making process. A higher importance could be given to the opinion of the executives, or some of the experts could be selectively excluded to highlight for example gender differences in decision-making.

The result of the aggregation is graphically shown in Fig. 9.11. Based on the answers of a group of experts, the average vector is considered for the multi-criteria analysis in mDSS. Overall, in this example the highest importance is assigned to the quantitative indicators.

By using the approach *Ordered Weighted Averages*, a different level of risk aversion can be explored. After normalizing and weighing the values of the analysis matrix into the evaluation matrix, the OWA approach sorts the resulting values for each option, from the lowest (criteria that have poor performance and/or low weights) to the highest (criteria with excellent performance for the option under consideration and/or high weight).

A risk-averse or pessimistic decision-maker will tend to focus on the first indicators with the poorest performance, assigning them great importance for the final decision. On the contrary, an optimistic decision-maker will want to highlight only the excellent performance of the best indicators. Other decision-makers can have a more risk-neutral attitude giving importance to both better and worse performances, neglecting to consider the criteria with average performance. The result of the OWA analysis for a neutral risk type as shown in Fig. 9.12. Alternative 3 is the preferable one if we assume that the decision-maker is risk-neutral (C-shaped OWA transformation function), but the overall ranking changes based on the decision-maker's

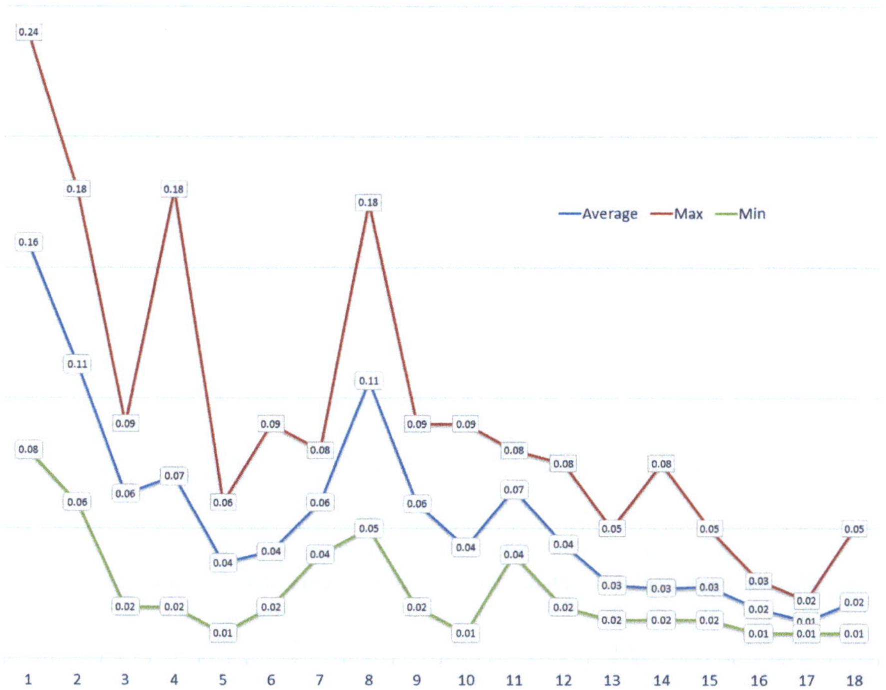

Fig. 9.11 Aggregation of importance weights

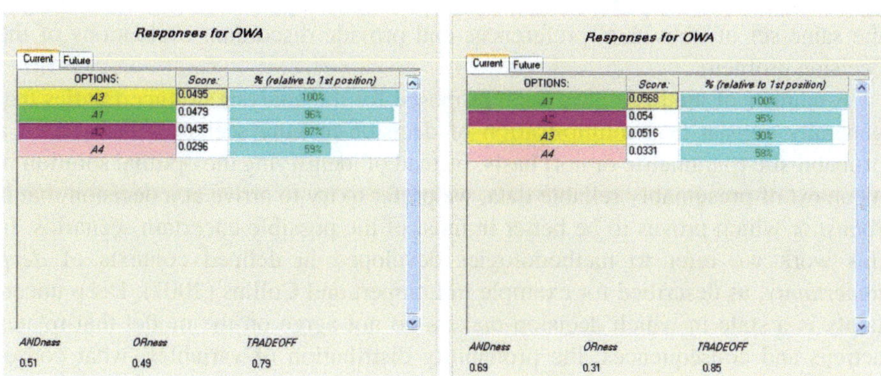

Fig. 9.12 Ranking after OWA—risk-neutral vs. risk-averse decision-maker

risk attitude, making the solution A1 the winning option and A3 the third best choice for risk-averse decision-makers.

The histograms in Fig. 9.13 provide further insight into the composition of the total score by providing a graphic idea of the contribution of each indicator.

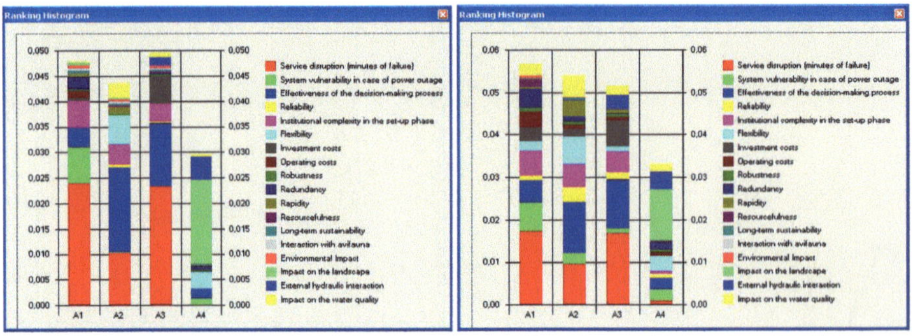

Fig. 9.13 Ranking histogram of risk-neutral vs. risk-averse decision-makers

Uncertainty Analysis (Step 8) and Evaluation of Options (Step 9)

The approach adopted with the classic MCDA represents what is normally done when one has to choose between alternatives considering also the trade-offs between evaluation criteria. The shortcoming of this approach is that it treats all data as certain, from the results of the hydraulic model to the estimated climatic and economic data and the scores provided by the experts.

As already mentioned in the methodological section, uncertainty can be divided into two types according to Walker et al. (2003): *aleatory* uncertainty, when estimates are made on future conditions that could undergo changes that are difficult to predict and to influence, and *epistemological* uncertainty, when experts do not share the same set of individual preferences and provide discordant evaluations of the decision problem.

A number of methods have been proposed for the analysis of uncertainties that generally consist in a multiplication of decision-making scenarios defined on a probabilistic (parametric or not) basis. Instead of identifying the optimal solution in a context of presumably reliable data, we prefer to try to arrive at a decision that is *robust* or which proves to be better in most of the possible uncertain scenarios. In this work we refer to methodologies developed in defined contexts of *deep uncertainty*, as described for example in Lempert and Collins (2007). Deep uncertainty is a state in which decision-makers do not agree on the model that relates actions and consequences, the probability distribution of variables, what consequences to consider and their relative importance.[6] To tackle this problem, we propose an intermediate solution: rather than working with a level of total uncertainty, we focused on plausible data intervals, even without assumptions about probability distributions. In this way reasonable alternative scenarios can be simulated, and it is

[6]Definition of *deep uncertainty* from the *Society for Decision Making Under Deep Uncertainty* (www.deepuncertainty.org).

Table 9.1 Robustness analysis OWA

	Option 1	Option 2	Option 3	Option 4
OWA_RA	48%	52%	0%	0%
OWA_RN	15%	38%	47%	0%
OWA_RT	15%	76%	9%	0%

possible to counterfactually verify whether a certain option prevails over others in many parallel "worlds".

An effective and at the same time intuitive method of robustness analysis are CART models (*Classification and Regression Tree*). These are decision tree models in which the sources of uncertainty are at the nodes and the emerging options at the final leaves.

To conduct the robustness analysis, we have to introduce some fuzziness around the deterministic values to assess whether small variation of either quantitative or qualitative indicator scores might reverse the ranking of the option. To simulate a real decision-making situation where agreements and compromises are reached through argumentation and dialogue, we sampled the matrices 100 times randomly within the observed Likert values of the questionnaires. Also, the four quantitative criteria were allowed to move within a range of +/− 10% of the estimated value. We assessed the possible effects of different risk attitudes of the decision-maker who could be risk-averse, risk-neutral or risk-taking. The results of the robustness analysis are shown in Table 9.1: the risk-averse decision-maker will be better off choosing the second option in 56% of the simulated scenarios. Similarly, in case of risk neutrality (C-shaped OWA aggregator), in 48% of the cases, solution 3 would be the preferable one. Optimistic decision-makers that assign a low importance to low criterion scores would favour alternative 4 three out of four times. Option 4 is never chosen, in which case we can safely conclude that it is dominated by the other alternatives.

As we can see, the robustness results show a slightly different picture than that of the crisp classic multi-criteria analysis where the input data consists in average indicator scores. By unravelling the panorama of possibilities, the robustness analysis shows that further assessments are necessary.

For a sensitivity analysis, we developed a CART algorithm in MATLAB/Simulink, which constructs a node for each indicator score that could possibly change the final choice. Below and above the limit values at the node, a reversal occurs in the ranking of the options. Figure 9.14 depicts the results of the CART analysis for the risk neutral decision-maker.

This tool is useful to assess whether indicator scores should be scrutinized more in depth. For example, if the value of criterion 6 (*flexibility*) exceeds the threshold value of 3.545, then either option 2 or option 3 will be the best alternatives, but option 1 is not a candidate anymore. To the left of the top C6 node, there is a wide scenario of possibilities where the choice of the best option depends on the limit values of the indicators at the nodes.

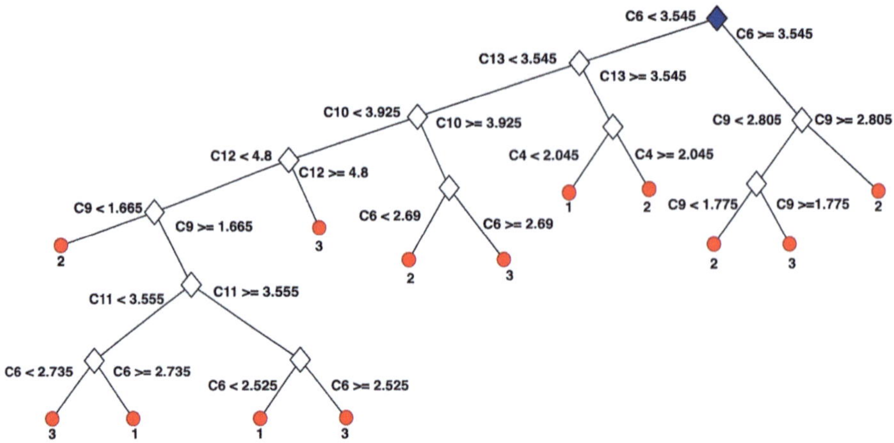

Fig. 9.14 CART analysis for risk-neutral decision-makers

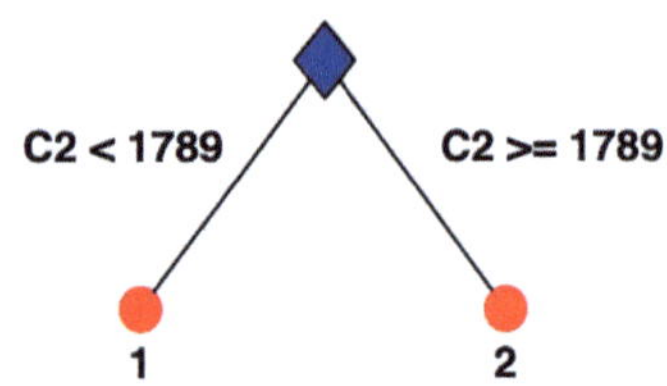

Fig. 9.15 CART analysis for risk-averse decision-makers

Similarly, for risk-averse decision-makers, Fig. 9.15 shows that the final ranking is sensitive to just one indicator, which is *system vulnerability in case of power outage*. Expert evaluations should therefore reach a firm consensus[7] for critical indicators in order to make sure that the chosen option is not at risk of being overturned by another one.

In our simplified example, we concentrated on current and future climate outlooks, risk attitudes[8] of the decision-makers, and on plausible as opposed to mean values of the indicator scores. But based on the type of decision problem, a variety of further robustness analyses can be conducted to assess whether the ranking stays unaltered. Examples are counterfactual analyses with different types of climatic and economic outlooks, alternative ways of weighing indicator scores and subgroup analyses of expert evaluations to reflect the decision power structure within the organization (executives vs. technical staff).

[7]There are several techniques to reach group consensus, most notably the Delphi method and the nominal group technique (Delbecq et al. 1975).

[8]CART simulation for risk takers available upon request.

Conclusions

Deep uncertainty means that there is a general disagreement about how the future will look like, about probability distributions and about parameter values. Hence, instead of following the traditional path of summing discrete criterion values, decision-making under these premises requires lying out all the plausible conditions under which an option emerges. Once the global panorama has been defined in that sense, the process continues recursively by looking into the likelihood of each single condition being met and by defining the assumptions that have to hold. The big picture can be created by using a CART (Classification and Regression Tree) algorithm as one of the possible computer-assisted descriptive analytics that identifies trigger indicator score values that make one option preferable to another one. The CART also supports decision-makers as it provides a familiar tree shape visualization outcome. With the aim to simplify the analysis and to provide decision-makers with a more intuitive presentation of the results, we develop a tool that represents an intermediate solution between traditional discrete top-to-bottom multi-criteria analysis and analysis under deep uncertainty.

The decision in our case study is to be made between four different infrastructural solutions (options, henceforth) to manage excess water due to extreme weather events. Before exploring the panorama of possible uncertainties, the starting point for the evaluation of the goodness of the options is assigning scores to a list of indicators that are relevant to the decision problem. The case of decision-making processes focused on flood risks is emblematic (Mastrandrea et al. 2011). The frequency and intensity of extreme events (e.g. floods, heatwaves, tropical cyclones and droughts) are rising globally. As a result, the economic damage as well as the number of people at risk have increased significantly, and these numbers are expected to grow further in the coming decades.

Decision-makers are interested in knowing how far unknown future events led by various related or isolated factors will influence their ability to adapt and cope with the negative consequences that may arise. After identifying the potential impact of the consequences, the main concern is to identify effective risk reduction solutions (*Risk Reduction Measures*, RRM). The selection process between a set of candidate solutions and an effective implementation depends, of course, on the specific objectives and constraints.

References

Ben-Haim, Y. (2004). Uncertainty, probability and information-gaps. *Reliability Engineering and System Safety, 85*(1–3), 249–266.

Bryant, B. P., & Lempert, R. J. (2010). Technological Forecasting & Social Change Thinking inside the box : A participatory , computer-assisted approach to scenario discovery. *Technological Forecasting & Social Change, 77*(1), 34–49. Elsevier Inc.

Darwiche, A. (2009). *Modeling and reasoning with Bayesian networks*. Cambridge: Cambridge University Press.

Delbecq, A. L., de Ven, A. H., Gustafson, D. H., & others. (1975). *Group techniques for program planning: A guide to nominal group and Delphi processes*. Scott, Foresman Glenview, IL.

European Commission. (2014). *SWD(2013) 137 – Adapting infrastructure to climate change*.

Fourali, C. (1997). Using fuzzy logic in educational measurement: The case of portfolio assessment. *Evaluation & Research in Education, 11*(3), 129–148.

Georgakakos, A. P. (2007). *Decision support systems for integrated water resources management with an application to the Nile basin*. Amsterdam: Elsevier.

Giupponi, C. (2007). Decision support systems for implementing the European water framework directive: The MULINO approach. *Environmental Modelling & Software, 22*(2), 248–258. Elsevier.

Giupponi, C. (2014). Decision support for mainstreaming climate change adaptation in water resources management. *Water Resources Management, 28*(13), 4795–4808.

Giupponi, C., Mysiak, J., Fassio, A., & Cogan, V. (2004). MULINO-DSS: A computer tool for sustainable use of water resources at the catchment scale. *Mathematics and Computers in Simulation, 64*(1), 13–24.

Giupponi, C., Sgobbi, A., Mysiak, J., Camera, R., & Fassio, A. (2008). NETSYMOD-an integrated approach for water resources management. In *Integrated Water Management* (pp. 69–93). Dordrecht: Springer.

Gorry, A. G., & Scott-Morton, N. (1971). A framework for management information systems. *Sloan Management Review, 12*, 55–70. MIT.

Groves, D. G., & Lempert, R. J. (2007). A new analytic method for finding policy-relevant scenarios. *Global Environmental Change, 17*(1), 73–85.

Haasnoot, M., van Deursen, W. P. A., Guillaume, J. H. A., Kwakkel, J. H., van Beek, E., & Middelkoop, H. (2014). Fit for purpose? Building and evaluating a fast, integrated model for exploring water policy pathways. *Environmental Modelling and Software, 60*, 99–120. Elsevier Ltd.

Herman, J. D., Reed, P. M., Zeff, H. B., & Characklis, G. W. (2015). How should robustness be defined for water systems planning under change? *Journal of Water Resources Planning and Management, 141*(10), 1–14.

IPCC. (2012). Managing the risks of extreme events and disasters to advance climate change adaptation: special report of the intergovernmental panel on climate change. In C. B. Field, V. Barros, T. . Stocker, D. Qin, D. J. Dokken, K. Ebi, M. D. Mastrandrea, et al. (Eds.), *Intergovernmental panel for climate change* (Intergover.). Cambridge, UK/New York: Cambridge University Press.

Kasprzyk, J. R., Nataraj, S., Reed, P. M., & Lempert, R. J. (2013). Many objective robust decision making for complex environmental systems undergoing change. *Environmental Modelling and Software, 42*, 55–71. Elsevier Ltd.

Keen, P. G. W., & Scott-Morton, M. S. (1978). *Decision support systems: An organizational perspective*. Reading, MA: Addison-Wesley Pub. Co.

Knight, F. H. (1921). *Risk, uncertainty, and profit*. New York: Boston, New York, Houghton Mifflin Company.

Lempert, R. J., & Collins, M. T. (2007). Managing the risk of uncertain threshold responses: Comparison of robust, optimum, and precautionary approaches. *Risk Analysis, 27*(4), 1009–1026.

Lempert, R. J., & Kalra, N. (2013). *Managing climate risks in developing countries with robust decision making*. Washington, DC: World Resources Report.

Lempert, R. J., Groves, D. G., Popper, S. W., & Bankes, S. C. (2006). A general, analytic method for generating robust strategies and narrative scenarios. *Management Science, 52*(4), 514–528.

Mahmoud, M., Liu, Y., Hartmann, H., Stewart, S., Wagener, T., Semmens, D., Stewart, R., et al. (2009). A formal framework for scenario development in support of environmental decision-making. *Environmental Modelling and Software, 24*(7), 798–808. Elsevier Ltd.

Mastrandrea, M. D., Mach, K. J., Plattner, G.-K., Edenhofer, O., Stocker, T. F., Field, C. B., Ebi, K. L., et al. (2011). The IPCC AR5 guidance note on consistent treatment of uncertainties: A common approach across the working groups. *Climatic Change, 108*(4), 675–691. Springer.

Rosenhead, J. (1980a). Planning under uncertainty: 1. The inflexibility of methodologies. *The Journal of the Operational Research Society, 31*(3), 209.

Rosenhead, J. (1980b). Planning under uncertainty: II . A methodology for robustness analysis. *The Journal of the Operational Research Society, 31*(4), 331–341.

Schoemaker, P. J. H. (1991). When and how to use scenario planning: A heuristic approach with illustration. *Journal of Forecasting, 10*(June 1990), 549–564.

Schoemaker, P. J. H. (1993). Multiple scenario planning. *Strategic Management Journal, 14*(January 1992).

Walker, W. E., Harremoës, P., Rotmans, J., van der Sluijs, J. P., van Asselt, M. B. A., Janssen, P., & Krayer von Krauss, M. P. (2003). Defining uncertainty: A conceptual basis for uncertainty management in model-based decision support. *Integrated Assessment, 4*(1), 5–17.

Yager, R. R. (1988). On ordered weighted averaging aggregation operators in multicriteria Decisionmaking. *IEEE Transactions on Systems, Man, and Cybernetics, 18*(1), 183–190.

Chapter 10
Spider: The Statistical Approach to Value Assignment Problem

Luigi Terruzzi

Abstract In the era of "Big Data," with the advent of the artificial intelligence algorithms, a different type of real estate valuation has found its own market niche: the AVM (automated valuation method), evolution of the simplest statistical evaluation. The Spider-based approach starts from the consideration that the value of a property is certainly influenced by innumerable variables, possibly coming from different databases, but these variables can be grouped into homogeneous categories, which we will call vectors.

Keywords Real estate evaluation · Market comparison approach · Cost approach · Income approach · Big Data approach

Introduction

Real estate evaluation is the process by which an opinion on the quality or value of a property or land is developed. A real estate valuation consists of various activities that culminate in the expression of a market value.

Real estate sales often require the performance of a real estate valuation since they do not occur frequently enough to be able to rely on predefined price lists.

Regulation 575/2013/EU defines in Art. 4 Paragraph 1 point 76 the market value as:

> The estimated amount at which the property would be sold on the valuation date in a transaction between a seller and a buyer consenting to normal market conditions after adequate commercial promotion, in which both parties acted with knowledge of the facts, with prudence and without being subjected to constraints.

L. Terruzzi (✉)
E&W, IT, Milan, Italy
e-mail: luigi.terruzzi@eaglewise.it

© Springer Nature Switzerland AG 2019
F. Cecconi, M. Campennì (eds.), *Information and Communication Technologies (ICT) in Economic Modeling*, Computational Social Sciences,
https://doi.org/10.1007/978-3-030-22605-3_10

The definition of the Open Market Value (OMV) therefore defines the value of the property substantially as the market price in which the property could be sold, after an adequate marketing period, between two consenting parties.

The market value is a representation of the exchange value or the advantageous price that could reasonably be obtained on the market if it were put up for sale on the free market at the date of the valuation in circumstances that meet defined criteria.

The market value is estimated by applying procedures and valuation methods related to the type, condition of the property, and the most probable circumstances in which this property would be sold on the free market.

The procedures and methods used to determine the market value must reflect the situations, data, and information of the real estate market. The choice of the valuation method is linked to the availability of data, the circumstances of the market, and the conditions of the property to be assessed.

There are different methods for calculating the market value of a property. Each of these methods can be applied to any building, but, as mentioned, for each type of property and at different market conditions, it is good to apply different methodologies.

The Comparison Method

The comparison method (MCA – market comparison approach) is based on the principle that in a reasonably liquid market, that are always properties that are being marketed, or have recently been sold, similar to the one being valued. Under these conditions, the market comparison method is the most direct, conclusive, and documented method for evaluating a property.

The OMV is determined by calculating the average of the buying and selling values of "comparable" properties, values obtained through an adequate market survey carried out locally by the real estate appraiser, and, usually, subsequently validated by an "auditor" structure (internal technicians who with the help of Borsini Immobiliari verify the result).

The "comparable" buildings are real estate units having characteristics similar to the unit subject to appraisal by area, degree of finishing, size and number of rooms, floors, exposure, brightness, etc. The market value also takes into account any other factors, visible and/or foreseeable at the time of the estimate, which could affect the value of the asset in question in the future. The evaluation is conducted by referring to the current destinationn of use of the asset, considering however also the impact on the value given by any alternative uses, if hypothetically allowed and convenient.

The basic assumption of the method is that it is possible to identify on the market a sufficient number of "similar" properties to the subject of the analysis ("subject") so as to be able to assume the average of the sales prices of the latter as the price of reference for the subject.

The results are weighted in relation to the marginal prices (see below) of the quantitative and qualitative characteristics of the properties taken in comparison with the one being valued.

It is worthwhile to briefly dwell on the concept of "similar" (the theme will be taken up later): given the qualitative variety of real estate (again, the subjective characteristics of the property influence its price), the hypothesis that there may be on the market a statistically significant number of properties perfectly representative of the subject, which are the object of a sale when the valuation must be carried out, or have been for a few months, represents a strong hypothesis, even in a very liquid market.

We briefly mention, for completeness, the other methods with which property valuations can be carried out, where the type of property, the purpose of the valuation (industrial, technical), or the scarce liquidity of the market do not suggest a valuation based on MCA.

The Cost Method (Cost Approach)

It provides an indication of the value based on the economic principle that the buyer will not pay for the property more than the cost for the purchase of a property of equal utility, by purchase or construction, including the land cost. It should be emphasized that with this method, it is then necessary to devalue the "just-built" property to take into account the age of the property being assessed. By value at the cost of reconstruction, we mean the cost necessary for the reconstruction of the building, in total respect of the surfaces, the volumes of the typological and architectural characteristics and the level of finishes of the individual real estate portions. The value is determined by applying a unitary parametric cost to the built-up area of the building.

Direct Capitalization Method (Income Approach)

The valuation is based on the ability of the assets to produce income. Direct capitalization transforms the income of a property into a market value through the use of a capitalization rate considered perpetual; this rate expresses the property's profitability in qualitative terms. The value of the property, therefore, corresponds to the present value of a postponed perpetual income and is obtained from the ratio between the rent and the appropriate capitalization rate.

Transformation Value

The current market value of an asset liable to change is determined by the difference in the market value of the transformed asset and the cost of the transformation itself. This method, is used in the case of assets that can be transformed or are currently being transformed. The value that is identified is given by the difference between the most probable market value of the transformed asset and the sum of all the most probable costs of the factors that will intervene in the transformation of the good itself.

Once the valuation method has been chosen, there three different ways to preform the appraisal: Full Appraisal (which is performed visiting the asset inside, room by room), Drive By Appraisal (performed viewing the asset outside), Desk Top appraisal (made in the office, without actually seeing the asset).

Full Appraisal

It is carried out with an internal and external inspection of the assets to detect their degree of functionality and efficiency, with analysis of the local economic landscape, with particular attention to the real estate market of reference, the orientation of the operators, the transactions carried out, the offer present, and at the time of absorption. Furthermore, the urban planning-cadastral compliance analysis of the asset is carried out, which highlights the presence of any discrepancies found during the inspection. In the event of discrepancies, the breakdown of the costs of placing the original plant and/or the evidence of the authorizations necessary for the asset to be transferable are indicated. The state of the places is also reported by a careful photo service to every single unit. The evaluation of each individual asset is carried out through a careful examination of the asset and the documentation that interests it, analyzing the elements useful for the purpose such as:

- Internal examination of the assets to detect their degree of functionality and efficiency, to be compared with those of recently bought properties (having similar characteristics in terms of finish and size) found on the market, for a more correct application of comparative values. This status of places also results from the photographic equipment made for each individual property unit and attached to the appraisal document (except for the desktop appraisals).
- Significant measurements, on a sample basis, of some rooms for each individual asset subject to assessment, in order to verify any discrepancies between the state of the places and the cadastral plans provided, often on a non-faithful scale.
- Analysis of the local economic landscape, with particular attention to the real estate market, the sensitivity of the operators, the transactions carried out, the present offer and the time taken for absorption, and the potential of the good and of alternative uses (HBU).

Drive-By Appraisal

The audit reports aim at determining the appropriateness of the commercial value, carried out by making an external visit of the building and on the basis of the data provided by the client, coming from documents such as preliminary sale deeds and/or previous purchase documents and/or original appraisal. A thorough market investigation is carried out, and a significant photographic kit is attached.

Desktop Expertise

The desktop appraisal are carried out in the office, without visiting the asset, on the basis of the documents provided, or gathered by the appraiser, using the own price listing of the company, which are based on the appraise expertise and previous valuations (known as "Borsini Immobiliari").

Regardless of the type of appraisal that must be carried out (FULL, DBy or DT), if the chosen methodology is that of the comparative (MCA), it will always be necessary to find a set of comparatives that will allow us to produce an estimate of the value of the property.

The Big Data Approach

In the era of "Big Data," with the advent of the artificial intelligence algorithms, a different type of real estate valuation has found its own market niche: the AVM (automated valuation method), evolution of the simplest statistical evaluation.

There are several reasons behind the fact that in these years, the AVM models are raising: on the one hand, the effects of the 2011 financial crisis, or rather the measures taken by the central banks to manage their effects, have created a request for evaluations of entire portfolios of properties to be carried out in a short timeframe and possibly at low costs (the banks found themselves in need of updating the valuations of the assets to guarantee their own mortgage loans, to assess the need to make provisions in the balance sheet).

Following this, when the banks decided to divest the entire portfolios of receivables, Both performing and non-performing, the investors interested in acquiring these portfolios found themselves in the need to evaluate the assets as collateral in a very short time, as the due diligence of a portfolio of credits often it takes place over a couple of weeks and with the lowest possible costs: more than ten potential investors are often involved in the preliminary (non-binding) phase of a competitive due diligence process while generally only four of them are clusters to the next phase.

Speed, economy and reliability: clearly a task to be carried out with a statistical-mathematical approach or, as we say today, with an AVM approach.

On the other hand, the great availability of data, available through the internet, has given impetus to the development of artificial intelligence algorithms whose predictions have become gradually more and more reliable.

In its simplest interpretation, a statistical-mathematical approach to real estate valuation means taking a database of real estate values and calculating the average value of the properties in a given geographical area. The result will be more precise the more granular the representation of the zones present in the database.

And of course, we are in statistics, the larger and more representative the sample of real estate in the database that meets the characteristics of the property will be.

The limit of this method is linked to the asset being valued: the property.

If we assume we can define a homogeneous geographical area from a socioeconomic point of view and consider two real estate units in two adjacent buildings. The values of these units will differ, for example, for the age of construction of the buildings. In the same building, the values will differ according to the surface of the building: the value per square meter will be smaller for a building of greater consistency. If they have homogeneous surfaces between them, they will differ for the floor within the same building: the upper floor tends to be more than one low floor. If they are on the same floor, they differ according to the side of the building the flat is facing: in Italy the building facing south is worth more than a building facing north; a double exposure is worth more than an apartment exposed on one side only of the building. Should all the features listed be equal; the finishes, the state of preservation, the energy efficiency, and so on are to be considered.

Each property can be considered unique.

A database containing real estate transactions, as detailed and deep as it might be can hardly include the analysis of all the features that have an effect on the value of the property.

In the execution of real estate evaluations with a "non-statistical" approach with the MCA method, described above, this limit is managed both through the competence of the assessor who carries out the inspection and through the principle of marginalization:

1. N characteristics of the asset to be valued are identified, considered significant variables to describe the asset being valued.
2. For the property subject to valuation (c.d. "subject"), the values of the N variables are measured.
3. The same characteristics are measured for the comparables buildings used as comparable.
4. The values of the characteristic variables are compared between subject and comparative to determine if the comparative transaction price can be used as a reference to determine the subject's valuation.

The comparison described in point 4 above works if there are transactions to be compared that have as their object properties perfectly identical to the subject. That is almost impossible.

It is therefore necessary to offset the value of each characteristic variable in order to compare "homogeneous" values. This is achieved by introducing a system of "weights," one for each of the N characteristics, which allows to "transform" each characteristic variable (c.d. "marginalization").

For example, suppose we have to evaluate an apartment of 100 square meters, located on the third floor of a recently built building, with a single southern exposure, in good condition.

As a comparison, a building of 80 square meters is identified, located on the second floor of a building built 25 years ago, with single southern exposure, in excellent condition, recently sold for 150,000. In order for the latter to be a good comparable for the former we need to "compensate" the fact that the surface is smaller by multiplying if for a parameter.

We then need to use a second parameter to weight the fact that our comparable is on the second floor, while the subject is on the third one. Then we will need to compensate the contribution that exposure and asset quality will provide to the valuation.

We leave any further details on the method of calculating the various parameters, which goes beyond the scope of this chapter, to the reading appraisal specific books. Here we are interested in highlighting how an approach based on a single parameter is largely insufficient to obtain a statistical evaluation algorithm that can calculate the value of a residential property automatically in a reliable way.

Even thinking of increasing the number of databases analyzed, in order to obtain results that have a much wider transaction base, will allow to obtain a more "authoritative" result as it is based on multiple sources but certainly not able to describe fully the object of the analysis.

However, the idea of using more than one database is not to be thrown away, but it is important to consider which databases to use.

The huge amount of data that is available today both for free through the Internet and through paid channels has shifted the focus from "where to retrieve information" to "what information to retrieve" and above all how to manage it. In fact, if it is unlikely to find a single database containing all the data necessary to fully evaluate a property, it is now equally easy to have the availability of different databases, containing specific information describing vertically individual sectors, in great detail.

For example, if we start from specific DBs where real estate values are classified according to use and location, or are aggregated by size (small, medium, large), we can consider adding the prices of real estate ads published on the Internet or data coming from notarial deeds of sale.

We can imagine supporting pure real estate transaction data bases with other databases containing information related to the quality of the area where the building is located: percentage of green areas, number of shops, location and distance from public transportation, from schools, cinemas, bank branches.

We can think of supporting a third line of investigation by analyzing the impact that the crime rates of the area can have on the value of the buildings or an analysis of the demographic growth/decrease trends of the area. If the asset being analyzed is a commercial activity, the data of the resident population may be significant, for a catering activity information about the number of people flowing daily in the area for work purposes could be relevant.

The facets can be the most imaginative: the availability of data today allows it. In the future we can only expect that it will increase further.

The focus has shifted from "where can I find information" to "how can I amalgamate information." In the evaluations carried out with classical techniques is the figure of the expert, a local technician who, thanks to his experience, knows the impact that all variables have on the value of the property. It is typically a non-codified competence, not measurable except through the years of experience that each individual technician can boast relative to a specific area. By the way, this is the reason why large companies that perform surveys throughout the country tend to have the most widespread network of technicians possible: not only to manage large volumes of appraisal but to be able to guarantee that each technician must cover small areas, in which he can boast a great knowledge.

A possible application solution of the problem could be the following: building an open tool, based on a variable number of sectorial databases which, through a learning algorithm (which we will call "fusion algorithm"), amalgamates the data by measuring the impact of each feature on the value of the property.

The algorithm will go, database by database, calculating the value of the property, evaluating the impact that the variables contained in each database, taken alone, have on the value itself.

The result of this analysis will be a vector of real estate values. It will then be the task of the fusion algorithm to calculate the final value of the property by also providing a series of votes for the reliability of each value expressed by the analysis of the individual databases.

To train the algorithm, it will be possible to use an additional database containing "actual" real estate valuations carried out by experts on buildings of various kinds, which are located as homogeneously as possible on the national territory.

These evaluations, by their nature, contain the maximum possible detail, the most varied real estate characteristics, and incorporate all the knowledge of the individual technicians of their reference territory. In fact, performing the analysis of such a database allows to train the algorithm as if investing years in direct training by the technicians.

A similar application was developed in Italy ("Spider" below), a typically low-liquid real estate market in the last decade (with the only exceptions of Milan and Rome), characterized by a high "difference" of the buildings. This is a context in which the assertion that there are no two identical buildings can be considered particularly valid: the probabilities of finding peculiarities between the building are high, even between buildings that were projected as identical.

The Spider-based approach starts from the consideration that the value of a property is certainly influenced by innumerable variables, possibly coming from different databases, but these variables can be grouped into homogeneous categories, which we will call vectors.

In particular, the Spider model rests on two specific vectors: the vector of the extrinsic variables (geographical vector) and the vector of intrinsic variables (vector of the characteristics).

The geographical vector collects those characteristics of the asset, which impact its value on the basis of the area in which it is located:

- The size of the municipality (number of inhabitants)
- The number of transactions held in the current period
- Closeness to means of transport/communication
- Gravitating population for work reasons
- Proximity to schools, supermarkets, and post offices

But also based on the context in which it is positioned, a shed placed in a residential area will have a much lower value than if it was placed in an industrial area no matter if, in the market of sheds is a very valuable shed.

The same shed will be worth more if placed near a motorway junction than in open country.

This set of variables affects all of the buildings in a homogeneous area (so-called OMI area) "[...] ie homogeneous market areas within which the historical data relating to the sale and rented property units are recorded. In particular, the OMI area is defined as a continuous portion of the municipal territory that reflects a homogeneous segment of the local real estate market, in which there is uniformity of appreciation for economic and socio-environmental conditions. Two perfectly identical buildings located in different OMI areas will have different values" (ref. Manual of the Quotes Database of the Real Estate Market Observatory).

The vector of intrinsic variables collects all those variables that impact the value of a property as it is made:

- Consistency
- Year of construction
- Energy performance
- Number of rooms
- Plan
- Etc.

The properties located in the same OMI area will have different values based on their intrinsic characteristics.

These two databases are joined by two other databases that can be retrieved via the web: real estate advertising values and OMI values.

Through the Internet it is possible to find a significant amount of real estate announcements related, more or less, to any area of the territory of which it is interested to carry out the evaluation.

Collecting all this information, cleaned by any duplication and historicizing them can provide not only a photograph of the real estate values of the area but also an estimate of the "liquidity" of the reference market of the property to be assessed, which in first analysis is directly proportional to the number of ads found in a given period of time. The historicization of data also allows to describe the c.d. "trend" of the market, allowing to highlight upward or downward trends very quickly.

By virtue of the speed with which the information is updated and the enormous amount of data available, the massive analysis of the websites, through "scraping" or "crawling" techniques, allows to intercept the manifestation of trends much more quickly than relying on other databases that, perhaps produced by more authoritative sources, are published semi-annually or annually and therefore require much longer time to draw price trends.

On the other hand, such a punctual analysis can be very disturbed by the "noise" present in the data and could highlight "micro trends" that do not then take on significance in an analysis even in the medium term, problems that can be solved quite easily through statistical techniques.

A very important element that characterizes the use of the data of the real estate announcements in the determination of the real estate values is the "negotiation discount" to apply to the data. The real estate announcement, almost by definition, does not represent the value to which the seller is willing to complete the sale. It is always advisable, if not necessary, to weigh the value present in each individual announcement by applying a discount percentage, generally variable in a range between 5% and 20%.

The actual value of the percentage to be applied depends on different variables. the liquidity of the reference market, the current price trend, and the palatability of a single property in the market and is, in a single assessment made by a technician, one of those information that strongly depends on the competence of the real estate technician itself.

The application, being able to carry out analyzes on market trends and its liquidity, can be calibrated to automatically apply discounts to the values of the ads and, working on very large amounts of data, it manages to obtain average values characterized by limited variances compared to the precise data produced by expert technicians.

With reference to the "noise" present in the data recovered from the Internet, it is possible to hypothesize the introduction of appropriately trained "semantic algorithms" so as to automatically intercept and discard those announcements that are not able to provide a positive contribution to the analysis. It is necessary to work in the field of semantics since often the real estate announcements have a set of information, few quantitative characteristics of the property, in tabular fields followed by a description of the property that often allows to assess whether the asset is, or not, to be included in the current analysis.

Therefore, from the Real Estate Advertising database we will be able to obtain average prices in an extremely granular way: even at street level and street number in particularly liquid markets, trends and liquidity of the market in which the property being analyzed are presented.

The OMI values database – Real Estate Market Observatory – is a database of real estate values published every 6 months by the Italian Revenue Agency and contains the values of the properties aggregated by homogeneous area (see above OMI Zones).

> "The database of real estate quotes (BDQ OMI) provides, on a six-monthly basis, for each limited area, a minimum interval – maximum of market and lease values, per unit area, by property type and state of maintenance and conservation 1 (commercial attractiveness for stores) (ref. Quotation Manual for the Real Estate Market Observatory)"

Below is an example of the information provided by the database consultation web interface.

Province	TR		
City	Marsciano		
Cadastral data	Via del Santo 12, 05101 234,43,65 Cd. 87.89.32		
Type	Ap. COD 23		
ID	Sup.	Min	Max
ID767.tr.mar.0023	120 Sq M	3.500 euro	3.800 euro

Semi-annual updates are provided free of charge to those who request it from the Land Agency (Agenzia del Territorio).

It is worth noting that many of the variables mentioned in the previous part of the chapter are ignored by the OMI quotatione: the Agency itself, in the aforementioned "Quotes Database of the Real Estate Market Observatory," specifies that "[...] given the particular complexity and heterogeneity that characterizes the real estate, the determination of the relative average prices can be subject to limits of representativeness. In this regard, it is necessary to specify how the OMI quotations cannot be understood as replacing the precise estimate of the value of the individual property, identifying a range of quotations in which the average value of real estate units in ordinary conditions is more likely to fall."

Despite these limitations, the OMI value database plays an important role in the application by offering a reference interval that allows the values produced by other databases to be calibrated.

Furthermore, by historicizing every single semester, it is possible to analyze the macro trends of the market, which go hand in hand with the data highlighted by the analysis of the property announcements.

At this point we have collected the data sources on which Spider is based:

Geographic DB
The geographical analysis extracts from the available databases the data of buildings actually assessed on the basis of a set of parameters that characterize the building based on "where it is positioned," such as the distance from the comparatives, the OMI microzone, the ISTAT coding, and the macro clusters (LTZ, Old Town Center).

The result of this processing is a vector of €/MQ values, each with its own "ranking" of reliability.

DB Property Features
The qualitative analysis proceeds by refining the outcome of the geographical analysis based on a set of "intrinsic" characteristics (up to 30 different parameters) of the building such as, by way of example, the consistency (or number of rooms), the floor, the number of bathrooms, the display, and the construction features.

The refinement takes place through a multivariate regression and produces as output a vector of €/Mq values.

DB OMI Values
For each property, which has been assigned the OMI code in the preliminary phase, the maximum and minimum values present are recorded.

WEB Crawling
The "crawling" algorithm scours the real estate ads on the web by classifying the ads on the basis of geographic features of the property by selecting comparatives falling in the same territorial area, or in areas with similar peculiarities, in which there is a substantial uniformity of appreciation for economic and socio-environmental conditions. The resulting data is properly discounted.

Also in this case the result is a vector of €/MQ values, each with its own "ranking" of reliability.

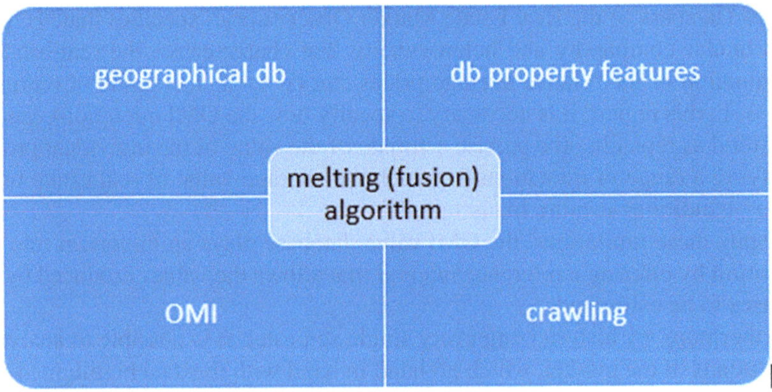

Fusion Algorithm

The values produced by the database analysis are analyzed by a learning algorithm which, using the vectors obtained in the statistical analysis phase and from the integration of the sources, generates a value in €/Mq and a percentage indication of the reliability of the data (PDA).

The PDA is a matrix

$$\begin{matrix} 85\% & \varepsilon_1 \\ 10\% & \varepsilon_2 \\ 5\% & \varepsilon_3 \end{matrix}$$

which indicates how big the error is with a probability of 85%, 10%, and 5%. For example,

$$\begin{matrix} 85\% & 150 \\ 10\% & 250 \\ 5\% & 500 \end{matrix}$$

It means with a probability of 85% the estimate on the value will be wrong of ±150 euros and so on for the next two values.

The result of the elaboration is at the same time a "fresh" datum, due to the origin of the real estate announcements available on the market, and "accurate," because it is weighed with the data coming from real estate appraisals carried out by experts of proven competence.

The result of the processing are the minimum-average-maximum values, of which the average value, is used to update the evaluation of the property.

The data sources incorporated by Spider came from the analysis of a very big data base, containing data distilled from more than 300.000 actual appraisals made by experienced technicians certified according to ISO17024 standard over the whole Italian territory.

Within this database there are the references of over one million transactions coming from comparative analysis, analysis of real market transactions, and interviews with real estate agencies.

The data are subjected to a statistical revaluation on a dynamic basis or based on the percentage increases in the values (+/−) over time highlighted by the analysis of the OMI data and the data produced by the crawling of Internet sites.

An application like Spider produces results that, in the field, have proven to be very consistent, producing results that oscillate in the order of +/− 10% compared to those desktop evaluations performed by professional technicians.

The availability of a wide database developed by professional technicians, which is updated daily through the data of new evaluations, is a fundamental element of the application since it adds a granularity to the quality of the data that is otherwise difficult to find with differen approaches.

On the other hand, it is necessary that the input data, i.e., the information relating to the property to be assessed, are equally granular.

An example can help to understand: if it is necessary to evaluate a property of which we know all the information necessary to compile a table similar to that already seen to describe the marginal, then it will be possible to appreciate the depth of the data set of which an application such as Spider can enjoy.

If, on the other hand, the information available on the property being evaluated is scarce, the algorithms will rely more on the average values than on the point values, will have to be made assumptions to populate the fields that would otherwise remain empty, some constraints will be relaxed by enlarging the meshes of the analysis, and the result will inevitably be less precise.

When evaluating entire portfolios of real estate, it is not infrequent to receive lists of addresses and stocks as input data, without further information. In these cases, lacking the full informational wealth of the characteristics database, the results will be more representative of the average values for those areas and those types of properties.

Chapter 11
Big Data for Fraud Detection

Vahid Mojtahed

Abstract Fraud is domain-specific, and there is no one-solution-fits-all method among fraud detection techniques. To make this chapter more specific and concrete, we provide examples concerning a common type of fraud which is food fraud. Food fraud has irreversible effects since it imposes risks to human life. The aim of this chapter is thus to present a conceptual and methodological solution for real-time fraud detection that can be implemented in the food sector by global food producers, regulatory bodies, or retailers but is generalizable to other domains.

Keywords Big data · Fraud detection · Anomaly detection · Clustering · Multivariate statistics

Introduction

Fraud is defined as making a false representation or failing to disclose relevant information or the abuse of position to receive a financial gain or misappropriate advantage (Cabinet Office 2014). Fraud activities, ranging from credit card and insurance fraud to food fraud, are very diverse depending on in which sector it is happening, but there are still common themes that enable us to use the same methodologies for detecting it. The victims of fraud can also vary from individuals or public sectors to businesses. The economic impacts of fraud go beyond the direct tangible costs that are often known as financial costs. Other less estimable dimensions of costs can be intangible (i.e., loss of reputation, loss of consumers trust, damage to consumers' health and loss of life, etc.) or indirect (i.e., loss of jobs, loss of tax income, etc.) which happens in the aftermath of fraud incidents through a

---

V. Mojtahed (✉)
Avanade Ltd., London, UK

Fera Science Ltd., National Agri-food Innovation Campus, Sand Hutton, York, UK
e-mail: Vahid.Mojtahed@avanade.com

© Springer Nature Switzerland AG 2019
F. Cecconi, M. Campennì (eds.), *Information and Communication Technologies (ICT) in Economic Modeling*, Computational Social Sciences,
https://doi.org/10.1007/978-3-030-22605-3_11

series of chain events (Button et al. 2009, 2014). Fraud is usually a concealed activity and therefore can go undetected. Despite this, studies have shown that fraud incident frequency is rising and on average 6.5% of income and expenditure of companies is lost due to fraud (Blakeborough and Giro Correira 2017; Gee 2018). The total cost of fraud is estimated to be £3.2 trillion globally (Gee 2018). It is of no surprise that detecting fraud can increase the competitive advantage of firms and increase the welfare of consumers; who else would pay more to cover the costs (i.e., fraud insurance premiums, disposal costs, etc.) passed to them by firms suffering from fraudulent or counterfeit products.

Detecting fraud is not an easy task as it requires information and knowledge about the nature of fraud (e.g., network intrusion, transactions and procurement, computer misuse, adulteration and dilution, mislabeling, false claim, etc.), as well as when, where, and how it has been committed. Fraud is the outcome of various types of changes in the status quo that creates lucrative opportunities for fraudsters to commit fraud. Depending on the context and domain, these changes can be in climate, regulations, social or economic conditions, etc. that render some industries more vulnerable to opportunistic fraudsters. By looking at all types of fraud, there is a vast literature on why fraud occurs in various sectors including food sector, which has significant impacts on human life, and there are some publicly known incidents such as the horsemeat scandal in 2013.

Fraud detection is an interdisciplinary concept and rooted in game theory, economics, statistics, social psychology, and criminology (Matsumura and Tucker 1992; Wilks and Zimbelman 2004). In criminology, Donald Cressey's fraud theory goes back to the 1950s and in short explains why fraudsters commit fraud, and it is conceptualized as "fraud triangle" which is proportional to the size of opportunity (Cressey 1950; Kassem and Higson 2012). From a game theory perspective, in order to catch a fraudster, you must be strategic and think like one and identify when and under which conditions a fraudster can maximize its profits. While criminology has more heuristic-based approaches (e.g., strategic thinking and auditing standards) to detect fraud, statistics and game theory employ analytical solutions such as data mining and machine learning techniques. The availability of various, high-volume, frequent datasets known as big data together with cost-effective predictive analytics such as machine learning algorithms can help us to overcome limitations in heuristic methods. Fraud detection in big data can change the current business models and develop more efficient ways to monitor and detect suspicious activities in markets, supply chains, financial transactions, insurance claims, etc. as part of the day-to-day risk mitigation strategies of businesses.

As mentioned before, fraud is often a concealed activity and goes undetected. As a result, there is limited amount of labeled data (i.e., fraud or no fraud) to be used for training analytical models that help us to learn from past events and predict future fraud incidents based on them. Even if the labeled data were abundant, we would suspect how useful they can be, in particular, when a new type of fraud emerges that has never been recorded before. In many domains, once a fraud incident is exposed to the public, the chances are that it will not occur again in the same way for some time or will manifest in a slightly different form and application; thus, they are not regularly repeated enough so that we could have records of their frequencies. For instance, in the agri-food domain, the type of fraud is always changing. One day it is melamine in milk (2010), the other day is an adulteration of beef with horsemeat (2013) or synthetic vanilla instead of natural vanilla (2016) (Hudson et al. 2016). In these circumstances, we cannot rely on limited historical data to predict a new type of fraud that has never been observed or recorded. However, if the kind of fraud is not changing over time and there are plenty of records of past events, we can build statistical models using the labeled data of fraud to detect or predict future incidents with success. This distinction will lead us to two different approaches in fraud detection analytical methodologies: (i) supervised machine learning and (ii) unsupervised machine learning. Our focus is on unsupervised machine learning for the various reasons that are explained in this chapter.

Fraud is domain-specific, and there is no one-solution-fits-all method among fraud detection techniques. To make this chapter more specific and concrete, we provide examples concerning a common type of fraud which is food fraud that often its consequences are heard in the news. Aside from that, food fraud has irreversible effects since it imposes risks to human life. The aim is thus to present a real-time fraud detection solution that can be implemented in the agri-food sector by global food producers, regulatory bodies, or retailers but its concepts and methods are generalizable to other domains. The benefits of the proposed methodology in comparison to other methods are as follows: (1) it is developed based on principals of big data; therefore, it is multivariate and can cope with frequent and diverse set of data for processing; and (2) it utilizes unsupervised machine learning algorithm, anomaly detection in particular, which enables it to automatically identify normal patterns in data and finds patterns that deviate from normal behavior or in other words the anomalies. This second feature is essential because it does not require the observations to be labeled with the outcome meaning fraudulent or nonfraudulent, which is a desirable feature for many real-time, online solutions. This means that the model is not anchored towards historical patterns of fraud observation.

This chapter is structured as follows. Section "Methodology" describes the Methodology behind big data fraud detection that is data intensive and uses anomaly detection algorithm. Section "Case Study and Results" demonstrates the application of this methodology to a constructed case study followed by section "Discussion and Conclusion", which concludes.

Methodology

Big Data and Fraud

Frauds are intentional actions with the motivation to gain economic gains (Spink and Moyer 2011; Tennyson 2008). The idea that we pursue in this chapter is: to detect fraud, we need to think like fraudsters and look at the factors that could influence the emerging size of fraud opportunity. The bigger is the size of a fraud opportunity, the higher is the likelihood of fraud to happen in that given time and circumstances. For instance, considering food fraud, many relevant factors have been suggested such as (i) global complexity of agri-food supply chain, (ii) high expenditure of consumers on food, and (iii) competitiveness of the food industry. It is essential to understand and disentangle the drivers of food fraud as signals that can be processed to derive insights. These drivers, in essence, are signaling the changes in the interconnected global or local markets. Examples of these drivers are global supply or demand shocks, local changes in food markets, exchange rates, regulations, or costs of local inputs such as energy, labor, raw materials, etc. We can group these signals into five main categories of characteristics of susceptible environment to food fraud based on the suggestions in the literature and the expert's inputs: (i) socio-economic characteristics, (ii) regulations and standard, (iii) industry or sector characteristics, (iv) commodities characteristics, and (v) climate, pests, and health issues. Under each of these main categories, we can identify many relevant signals taken from public data sources or propitiatory data warehouses of businesses. The set of all these signals creates what we call a "big data landscape" to which we can apply our machine learning algorithms for a continuous monitoring of any sudden changes. Note that not all these signals may be active during a fraud incident, but it is important to monitor them to maximize the likelihood of detecting future frauds.

Anomaly Detection

Frauds are considered to be rare events, and therefore data regarding fraud incidents are often scarce as only a small fraction of fraud incidents are detected and thus reported. Even if labeled data were readily available, the benefit of them is limited or questionable as the characteristics and nature of food fraud are rapidly changing over time, often rendering historical data unsuitable for future prediction. Fraudsters always tend to find innovative techniques and methods to commit their crime without being detected or caught, and chemical tests, which are used in the agri-food industry, are often lagging in catching up with them.

For the reason mentioned above, we aim to analyze the big data landscape using statistical methodologies combined with machine learning algorithms that are restricted to unsupervised learning. By unsupervised learning, we mean that there is no need for the data to be labeled by type of fraud incidents in order to be able to infer or predict future fraud incidents. This is in contrast to the supervised methodologies that rely on datasets consisting of a pair of inputs vectors and desired/undesired output values or labels. These types of unsupervised methods are particularly suitable for developing our intended fraud detection system since it requires minimum inputs from end-users for performing its task, and also it does not require any new training dataset providing further information.

Another important criterion for searching and selecting the right algorithm for our purpose is the capability of the algorithm to process high-dimensional and high-frequency data in a fast, efficient, and reliable manner. As mentioned earlier, we can have a data landscape that is on an hourly or daily basis, across many branches/regions/countries, and at a very fine granularity. This means that for a fraud detection system to be successful in identifying fraud, preferably in real time, it should be able to process a large amount of multivariate data not only comprised of main signals but also other ancillary datasets.

In fraud detection analysis, we should also be aware of the type of data. If the data points are related to each other (i.e., sequential data, spatial data, etc.), then they are ordered. In such cases such as time series (as one type of sequential data), we face the higher-order problem of identifying uncommon patterns beyond single outliers. For instance, fraud can span over several months, and we need to define which period is the benchmark as our normal pattern. Additionally, when dealing with real datasets, they are generally not multivariate normal with possible missing records. This imposes an additional challenge especially when ancillary data are not available to validate our findings.

The problem of outlier detection is well studied in statistics and data mining literature (Cerioli and Farcomeni 2011; Chandola et al. 2009; Filzmoser et al. 2005, 2008; Filzmoser and Hron 2008; Patcha and Park 2007; Riani et al. 2009). We focus on a family of candid techniques for identifying anomalies in a complex big data landscape when there is no clear indication between normal pattern and unusual activity that could signal fraudulent activity. The advanced analytics that we are using is based on a series of procedures that starts by reducing the dimensionality of big data. At the same time and during the process of reducing the dimensionality, we seek to identify a normal pattern in the collective signal data. All the data points that do not adhere to this normal pattern are then tested against a reference distribution to ensure that our method is robust and insensitive to the presence of big outliers that could distort the normal patterns of the data and show a false positive.

Since we are considering big data, the emphasis is on multivariate outlier and not necessarily identifying an extreme in any individual signal (Barnett and Lewis 1994). Outliers are data points coming from one or more different distributions, and extremes are instead values that are far away from the center but belonging to the same distribution. This is the reason that we have characterized the fraudulent activities as a result of collective deviations from the status quo or what is considered to be normal conditions. The collective set of signals could include correlated signals (e.g., prices of complementary or substitute commodities), or it could consist of signals that are uncorrelated (e.g., change in regulations and climate signals). For instance, simultaneous but small changes in a set of signals could result in an anomaly from fraudulent activities. One example of this is the horsemeat scandal in 2013 where there were changes in the import of horsemeat and the prices of horsemeat and beef together with the change in regulations (Guardian 2013).

Anomaly Types

Three types of anomalies can be identified by pointing to different causes. These are point anomalies, contextual anomalies, and collective anomalies (Chandola et al. 2009; Gogoi et al. 2010).

Point Anomaly Point anomaly is the simplest and the most widespread type of anomaly. It refers to an individual data point that is anomalous with respect to the rest of the data.

Contextual Anomaly A data point is a contextual anomaly if it is anomalous in a specific context. The context is brought about by the structure of the data and needs to be specified as part of the problem formulation (Wang et al. 2011). The anomalous behavior is determined using the values of the behavioral attributes within a specific context. A data point might be a contextual anomaly in a given context, but an identical data instance could be considered a normal data point in a different context (Lane and Brodley 1997). Another critical factor is the availability of the contextual attributes. In some cases, defining a context is straightforward, but in other cases, it might be difficult to apply such techniques as setting the context for the domain of application requires extra efforts to determine an appropriate context.

Collective Anomaly If a subset of related data points is anomalous with respect to the whole dataset, then they are referred to as a collective anomaly. In this type of anomaly, an individual data point in the subset may not be considered anomalous per se, but the joint observation of all data points in the subset is a signal of anomaly (Patcha and Park 2007).

It is worth to note that point anomalies can happen in any database, while collective anomalies occur in data sets with data points that are related to each other. Contextual anomalies, on the other hand, take place when a set of context attributes exist in the data. A point or collective anomaly can be a contextual anomaly whenever they are analyzed in a particular context. Therefore, a point anomaly detection problem or collective anomaly detection problem can be transformed into a contextual anomaly detection problem by adding a context to the analysis.

Anomaly Detection Methods

Anomaly detection refers to the problem of finding patterns in data that are not complying with expected behaviors. Such abnormal patterns are referred to as anomalies, outliers, discordant observations, surprises, or contaminants in the different application domain (Chandola et al. 2009). In the realm of food fraud detection, an abnormal pattern of trade for a given commodity from a given country in a given period can indicate a fraud incidence.

Various techniques such as classification, information theoretic, spectral, etc. have been proposed for anomaly detection and have been applied to different problem formulations. However, few of them are suitable for big data (high-frequency multivariate data) that has no labels. Below, we present an anomaly detection technique that combines clustering with multivariate statistics.

Distance and Outlier Detection

Given a particular data point in the big data, the key question is how rare or extreme is the data point compared to other data points. What is the distance of the data point from the center of the distribution? Also, how likely is that the data point will be that far from the center of the distribution?

To answer these questions for a univariate dataset, we can refer to the Euclidean distance, and rarity will be determined consequently. However, for big data, we use a different concept called Mahalanobis distance which enables us to detect multivariate outliers. We calculate the distance of a point $\vec{p}$ to the centroid $\vec{\alpha}$ using a non-Euclidean distance measure based on a weighted distance of z-scores as follows:

$$\boldsymbol{d} = \sqrt{\left(\frac{x-\mu_x}{\sigma_x}\right)^2 + \left(\frac{y-\mu_y}{\sigma_y}\right)^2} \tag{11.1}$$

This measure does not have the flaws of Euclidean distance; therefore (i) it is dimensionless, so we avoid problems of mixing different measurement units; (ii) it is rescaling the signals to standard deviation units so the problem of different scales for signals is evaded; and (iii) it gives less weight to more signal factors when it rescales the signals based on the standard deviations. We can derive the locus of all points having the same weighted distance from centroid by restructuring Equation (11.1):

$$\mathbf{1} = \left(\frac{x-\mu_x}{d\sigma_x}\right)^2 + \left(\frac{y-\mu_y}{d\sigma_y}\right)^2 \tag{11.2}$$

Equation (11.2) describes an ellipse centered at the centroid $\vec{\mu} = \left[\mu_x \ \mu_y\right]'$ and having axes of length $d\sigma_x$ and $d\sigma_y$. The iso-weighted distance contours demonstrates that points positioned along the same ellipse are considered to have the same statistical extremeness from the centroid even though their Euclidean distances can differ. The ellipses are interpreted as probability density contours. Equations (11.1) and (11.2) can be extended, without loss of generality, to consider several signals. In that, we have

$$d^2 = \sum \left(\frac{x_i - \mu_{x_i}}{\sigma_{x_i}} \right)^2 \tag{11.3}$$

where x_i, $i \in N$ is the set of signals in the big data landscape. The equation now describes a hyper-ellipsoid with an iso-weighted distance shell and is interpreted as a probability density shell. Equation (11.3) is the sum of z-scores of several variables, and it is also the equation of a chi-squared (x^2) probability distribution, which will be essential for assessing the extremeness and rarity of the weighted distances (Johnson and Wichern 2002). So far, we have assumed that all variables are uncorrelated and the covariance matrix is zero. This is an unrealistic assumption in the context of our analysis, and we will eliminate this limitation by using the distance metric developed by Mahalanobis (1936). It is worth to note that Mahalanobis distance could easily be generalized to the weighted distance between any two multivariate points and it does not need to be necessary between a point and the centroid.

Assume an ellipse in a bivariate scatterplot as shown in Fig. 11.1. All points on the ellipse are equidistant from the center of the ellipse and thus statistically speaking are equiprobable. A set of nested concentric ellipses each corresponding to one probability value are referred to as probability density ellipses.

In general, if we have multivariate data points $\vec{x} = \left[x_1, x_2, \ldots, x_p \right]^T$ and $\vec{y} = \left[y_1, y_2, \ldots, y_p \right]^T$ drawn from a set of p variables with $p \times p$ covariance matrix, then the Mahalanobis distance d_m between them is defined as

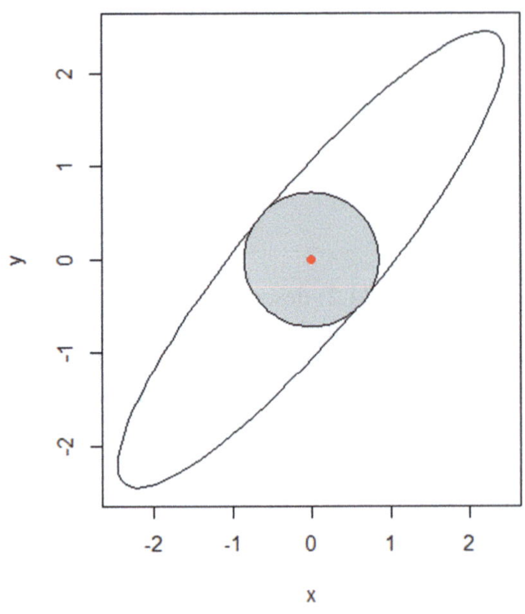

Fig. 11.1 *Mahalanobis distance*

$$d_m(\vec{x}-\vec{y})=\sqrt{(\vec{x}-\vec{y})^T S^{-1}(\vec{x}-\vec{y})} \tag{11.4}$$

If the underlying distribution of the p random variables is exactly multivariate normal with the $p \times p$ covariance matrix $\boldsymbol{\Sigma}$ and if $\vec{y}=\vec{\mu}$ the Mahalanobis distance d_m of a particular multivariate data point $\vec{x}$ from $\vec{\propto}$ is

$$d_m(\vec{x})=\sqrt{(\vec{x}-\vec{\mu})^T S^{-1}(\vec{x}-\vec{\mu})} \tag{11.5}$$

setting d_m to a constant c defines a hyper-ellipsoid with a centroid at $\vec{\propto}$. The shell of the ellipsoid is a probability density contour, and the probability associated with each c^2 shell follows a x^2distribution with p degrees of freedom. That is an ellipsoid of x values that satisfy the following equation:

$$(\vec{x}-\vec{\mu})^T S^{-1}(\vec{x}-\vec{\mu}) \le x_p^2(\alpha) \tag{11.6}$$

has probability$1-\alpha$.

Note that we are using S as the empirical or sample covariance matrix rather than $\boldsymbol{\Sigma}$ as the theoretical or population covariance matrix which gives us greater flexibility.

Mahalanobis distance has several advantages for being used to detect outliers such as:

- It provides numerical and graphical thresholds to identify the outliers.
- It is an alternative to econometric techniques when we don't have a value to predict such as the probability of fraud.
- It can predict unusual patterns within a multivariate observation.
- It has the capability to reduce the impact of outliers during searching for outliers.
- It allows using robust and independent values for centroid and covariance matrix.

The fact that the Mahalanobis distance is associated to the chi-squared probability distribution allows us to identify candid outliers based on thresholds coming from confidence levels and degrees of freedom. For instance, if we have four variables ($p = 4$) and want a significance level of $\alpha = 0.01$, the x^2 will be 13.28. This means that given Equation 11.6, we expect 1% of squared Mahalanobis distances to be higher than 13.28. Graphically this means that only 1% of points should lie outside of an ellipse whose contour is defined by

$$d_m=\sqrt{c^2}=\sqrt{x^2}=\sqrt{13.28}=3.64$$

Therefore, the square root of the critical value of x^2 can be our threshold for deciding which data points are outliers.

The Mahalanobis distance d_m is supposed to tell us how far a data point is from the center of a data cloud by also considering the shape of the cloud. However, this approach suffers from masking effects, which means that some of the outliers might not have a large d_m. This is because the location and scale parameters μ_x and S^{-1} are not robust. This means that a small number of outliers can attract the arithmetic mean and inflate the covariance matrix in their direction (Rosseeuw and Van Zomeren 1990). This can be puzzling because the Mahalanobis distance is supposed to detect outliers, but the same outliers can profoundly affect that distance (Filzmoser et al. 2005). Therefore, we need to replace them with robust estimates. By robust we mean resistance against the influence of the outliers. Once the Mahalanobis distance is estimated by robust procedures, we have reliable measures for recognizing outliers. Note that no outlier is deleted within these robust procedures, but the purpose is only to mitigate their impact on the identification of some of the smaller outliers that are masked in the presence of bigger ones known as leverage points.

The minimum covariance determinant (MCD) (Rousseeuw 1985) estimator is the most popular method in practice as it is computationally fast. The MCD estimator is calculated using a subset of the observations that minimizes the determinant of the sample covariance matrix. The location estimator (arithmetic mean) is the average of this subset of points. The scale parameter is proportional to the covariance matrix. Around 75% of the whole data is chosen for forming the subset that will be used to estimate the robust estimators (Filzmoser et al. 2005). This is a compromise between robustness and efficiency. The higher is the fraction of the data used for forming the subset, the lower is the breakdown. The breakdown is defined as the fraction of the outliers that, when exceeded, will cause the estimators to be biased.

Once the robust estimators is used for location and scale parameters in Equation (11.2), we arrive at the so-called robust distance d_r. Now, if the squared d_r for a data point is greater than, say x^2 of 0.99, the point can be declared an outlier. The threshold is somehow subjective and can be adjusted to the sample size. Moreover, there is no reason to fix this threshold and use it for every dataset (Filzmoser et al. 2005).

The chi-square plot is useful for visualizing the deviation of the data distribution from the multivariate normality in the tail. It is a common practice by some researchers to plot the ordered Mahalanobis distances and plot them against their corresponding chi-squared values (48,50). In this approach, the *i*th ranked Mahalanobis distance d_r^2 out of *N*, with cumulative probability $p = \frac{(i - 0.5)}{N}$ plotted against the value of $\boldsymbol{x}_p^2(\alpha)$ where p is the number of variables. A multivariate normal distribution will plot as a straight line from the origin (0,0) along the 45° line (Garrett 1989).

If a dataset is well behaving in a chi-square plot, the points are linearly distributed along the 45° line. When outliers are present, then the chi-square plot is not well behaved. In particular, with poly-populational data display, we will observe straight line segments on the chi-square plot. A useful outlier indicator is when some of the data points appear in the continuation of but separated from the main mass of the data displaying a gap or change in the slope of display of data.

Usually, the detection of the anomalies or outliers is done using a fixed threshold value coming from a given chi-square critical value corresponding to certain degrees of freedom and quantile. However, defining anomalies using this method is somewhat subjective because:

- The threshold has to be adjusted to the sample size.
- There is no reason to believe that a fixed threshold should be suitable for every dataset.

Therefore, a better method has been proposed based on adjusting the threshold to the dataset at hand. For instance Garrett 1989 has proposed plotting the squared robust Mahalanobis distances against the quantiles of the chi-square distribution and deleting the most extreme points until the remaining points follow a straight line. These removed points are the identified anomalies. The drawback of this method is that it is not automatic, and it requires user interaction and expertise of the analyst who can be subjective. Besides, for large datasets, it can be very time consuming and tedious.

The alternative objective and automatic solution are described by Filzmoser et al. (Filzmoser et al. 2005) as the following: Let $F_n(u)$ be the empirical distribution function of the squared robust Mahalanobis distances, and let $F(u)$ be the chi-square distribution function with p degrees of freedom. If the data were multivariate normal, the two distributions converge, and the tails of the two distributions can be compared together to detect outliers. The tails will be defined by $\delta = x^2_{p,1-\alpha}$ for a given small α and

$$p_n(\delta) = \sup\left|\left(F(u) - F_n(u)\right)\right|$$

In this way, $p_n(\delta)$ measures the departure of the empirical from the theoretical distribution only in the tails, defined by the value of δ. $p_n(\delta)$ will be considered as the indirect proxy of the anomaly in the sample. The next step is to define a critical value p_{crit} which helps us to distinguish outliers/anomalies from the extremes in the distribution. The measure of the outliers in the sample is then defined by

$$\alpha_n(\delta) = \begin{cases} 0 & \textit{if } p_n(\delta) \le p_{crit}(\delta,n,p), \\ p_n(\delta) & \textit{if } p_n(\delta) > p_{crit}(\delta,n,p). \end{cases}$$

The threshold value is then determined as $c_n(\delta) = G_n^{-1}\left(1 - \alpha_n(\delta)\right)$. The critical value p_{crit} is derived from the simulation for different sample sizes n and different dimensions.

Case Study and Results

We construct a case study similar to the events of horsemeat scandal in 2013 to demonstrate how fraud detection can be applied to detect food fraud. Let us assume that we have the prices and bilaterally traded quantities of over ten different types of meat (e.g., bovine, swine, mule, poultry, etc.) in a period of 48 time-steps (say monthly). In this case, fraud is happening through adulterating meat (i.e., using a cheaper, lower-quality substitute mixed with higher-quality, more expensive type) which is a common type of fraud in the food industry. We have done this analysis in two steps using R (R Core Team 2014) and the package *mvoutlier* (Filzmoser et al. 2005). At the first step, we analyzed prices and quantities at an aggregate level summing up all quantities of various types of meat and used the average price of meat together with the ten disaggregated types. In the second step, we analyzed prices and quantities of all ten types of meat separately without adding the aggregated

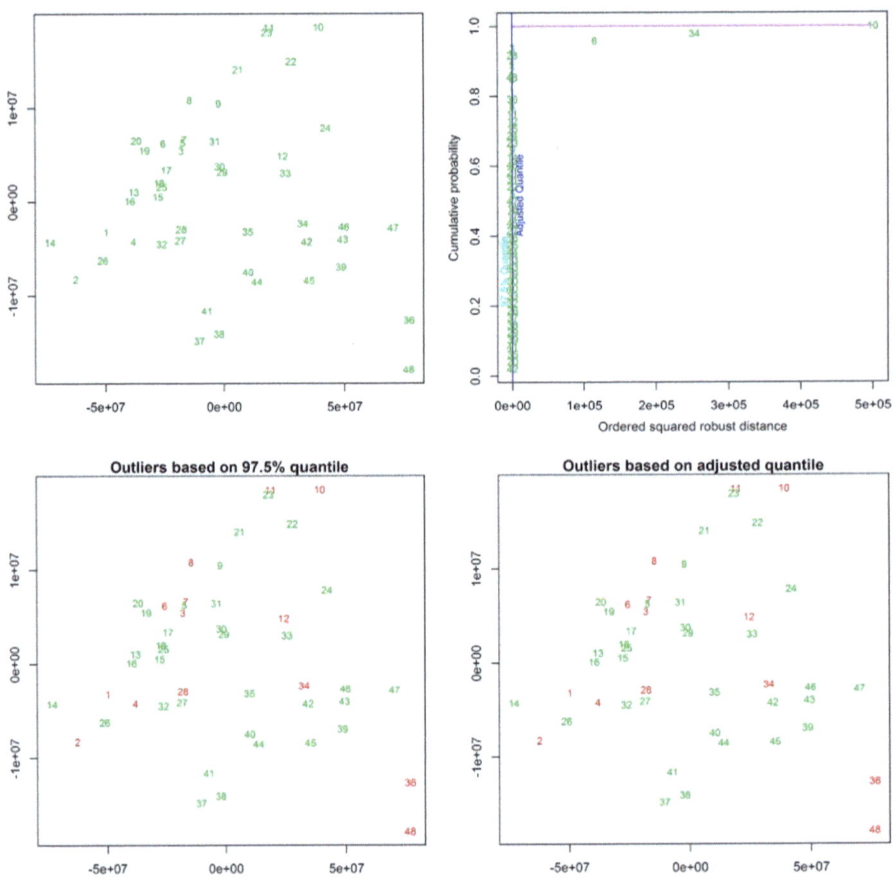

Fig. 11.2 Anomalies in the aggregated quantities of meat given 97.5% confidence

quantity and average price to the analysis. The purpose of this two-step analysis is to save time. We can always start with a big data landscape and put all signals together for the analysis. Once we detect an anomaly, we can zoom in and pinpoint the exact reason behind the found anomaly.

We are depicting the distances of the 11-dimensional space of monthly records in the top left-hand side of Fig. 11.2. The top-right figure in Fig. 11.2 plots the robust distances (d_r^2) of traded meat (all categories at the same time) against the empirical distribution of the distance $\left(d_m^2\right)$ shaped by the number-month of the data. Moreover, our reference distribution function is plotted along with two vertical lines demonstrating the specified distribution quantiles and the adjusted quantile. The adjusted quantile divides the anomalies from the normal pattern of the data. The two bottom figures show the outliers detected based on the specified quantiles (97.5%) of the reference distribution and the adjusted quantile.

In Fig. 11.3, we plot the multivariate anomalies that we found in the first step and plot each category of meat type in a one-dimensional scatter plot. The anomalies are marked using a combination of symbols and colors. Based on robust distances, the cross means a big anomaly and the circle indicates a small anomaly in terms of magnitude, and according to the Euclidean distances, red means a big value for anomaly and blue means a small value (Filzmoser et al. 2005).

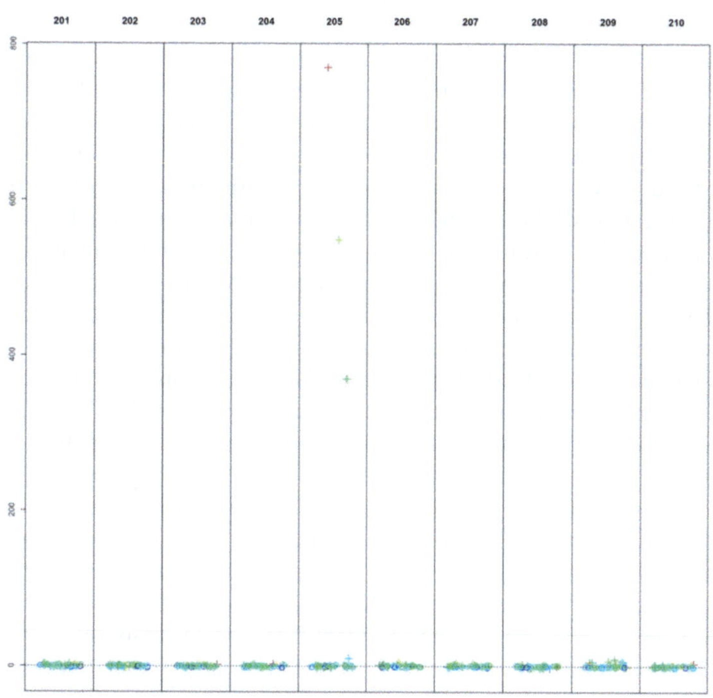

Fig. 11.3 Anomalies in trade of disaggregated meats

As seen in Fig. 11.3, all the three anomalies are associated with the trade in commodity 205 which we constructed to be the commodity with lower price used for adulteration. The most significant anomaly is the one that relates to month 10 of the dataset. This is related to the time-step that the quantity of the traded commodity has relatively increased. The important context to this anomaly is that if the quantity of any of these meat categories increases in such a short period of time, it is alarming, since the consumption can never increase over night! Therefore, the identified anomaly under this context looks suspicious and points to a fraudulent activity.

Discussion and Conclusion

Slight changes in the status quo could create opportunities for fraudsters anywhere in the world to commit fraud. It is therefore essential to continually monitor the underlying circumstances and look for a set of changes that collectively could increase the susceptibility of a domain to fraudulent activities. Not long time ago, this was a challenging task, which would require many resources for systematically collecting and processing the data and providing a meaningful synthesis of fluctuations in the data. However, nowadays, it is prevalent to hear that the information has gone from rare to abundant. The availability of new sources of information has brought us new benefits in terms of combatting crime by using sophisticated algorithms applied to big data.

Even if labeled data are not available, we can still detect fraud deploying unsupervised machine learning algorithms. Anomaly detection algorithms determines anomalies in the signals, collectively or individually, which are associated to potential fraud opportunities. It is very difficult to establish the ground truth since fraud is a concealed activity. However, by studying the context under which fraud happens, we can establish the link between these anomalies and fraud incidents. For instance, increased price of a rare commodity which is highly demanded, multiple transaction in a short period of time, increase in import of a commodity that its demand has not changed, and transactions in an irregular time period such as credit card transactions in the night can be all red-flagged given the context.

The constructed case study that was provided is a simple example of how the anomaly detection can be applied to a set of multivariate signals. This approach can be coupled with other techniques to consider spatial or temporal patterns of fraud. We briefly describe two examples of it here: (a) using concepts from social network theory combined with anomaly detection algorithms, we can detect anomalies across many countries and commodities which in turn reinforce the raised red-flagged anomalies indicating systematic and organized attempts are being made to commit fraud globally, and (b) by tweaking the algorithm and applying it to the temporal slices of data, we are able to identify the right temporal context within which the anomalies can reveal themselves. This is done by considering a wider set of patterns in data which helps us to rule out anomalies that were detected because of the wrong context. For instance, agricultural productions follow a seasonality pattern which can takes months to years. We can always slice the data into overlapping periods to see if a certain anomaly is being repeated in every time-slice or is only present in a certain time-slice and interpret that as a fraudulent activity.

Last but not least, incorporating additional non-numerical datasets including sensitive data that are not necessarily publicly available could point us in the direction of the causes of the fraud. And knowing the underlying reasons of anomalies will consequently reduce the likelihoods of false alarms and improve our predictability.

References

Barnett, V., & Lewis, T. (1994). *Outliers in statistical data*. 3rd edition, John Wiley & Sons, Chichester, UK, (pp. 584), ISBN 0-471-93094-6.

Blakeborough, L., & Giro Correira, S. (2017). *The scale and nature of fraud: A review of evidence*. ISBN 978-1-78655-682-0 (evidence review undertaken by Home Office Analysis and Insight to bring together what is known about the scale and nature of fraud affecting individuals and businesses in the UK)

Button, M., Lewis, C., & Tapley, J. (2009). *Fraud typologies and the victims of fraud: literature review*. London: National Fraud Authority, 40 p.

Button, M., Lewis, C., & Tapley, J. (2014). Not a victimless crime: The impact of fraud on individual victims and their families. *Security Journal, 27*(1), 36–54.

Cabinet Office. (2014). Common areas of spend, Fraud, error and debt, Standard Definition v2.1. Retrieved from http://www.gov.uk/government/uploads/system/uploads/attachment_data/file/340578/CAS-FED-Guidance-version-2.1-July-2014_P1.pdf

Cerioli, A., & Farcomeni, A. (2011). Error rates for multivariate outlier detection. *Computational Statistics & Data Analysis, 55*(1), 544–553.

Chandola, V., Banerjee, A., & Kumar, V. (2009). Anomaly detection: A survey. *ACM Computing Surveys (CSUR), 41*(3), 15.

Cressey, D. R. (1950). The criminal violation of financial trust. *American Sociological Review, 15*(6), 738–743.

Filzmoser, P., & Hron, K. (2008). Outlier detection for compositional data using robust methods. *Mathematical Geosciences, 40*(3), 233–248.

Filzmoser, P., Garrett, R. G., & Reimann, C. (2005). Multivariate outlier detection in exploration geochemistry. *Computers & Geosciences, 31*(5), 579–587.

Filzmoser, P., Maronna, R., & Werner, M. (2008). Outlier identification in high dimensions. *Computational Statistics & Data Analysis, 52*(3), 1694–1711.

Garrett, R. G. (1989). The chi-square plot: A tool for multivariate outlier recognition. *Journal of Geochemical Exploration, 32*(1–3), 319–341.

Gee, J. (2018). The financial cost of fraud. Retrieved from https://www.crowe.com/uk/croweuk/insights/financial-cost-of-fraud-2018

Gogoi, P., Borah, B., & Bhattacharyya, D. K. (2010). Anomaly detection analysis of intrusion data using supervised & unsupervised approach. *Journal of Convergence Information Technology, 5*(1), 95–110.

Guardian, T. (2013). Horsemeat scandal blamed on European meat regulation changes. *The Guardian*. Retrieved from https://www.theguardian.com/environment/2013/feb/12/horsemeat-scandal-european-regulation-changes

Hudson, A., Thomas, M., & Brereton, P. (2016). Food incidents: Lessons from the past and anticipating the future. *New Food, 19*, 35–39.

Johnson, R. A., & Wichern, D. W. (2002). *Applied multivariate statistical analysis* (Vol. 5). Upper Saddle River, NJ: Prentice Hall.

Kassem, R., & Higson, A. (2012). The new fraud triangle model. *Journal of Emerging Trends in Economics and Management Sciences, 3*(3), 191.

Lane, T., & Brodley, C. E. (1997). Sequence matching and learning in anomaly detection for computer security. In *AAAI Workshop: AI Approaches to Fraud Detection and Risk Management*, pp. 43–49.

Matsumura, E. M., & Tucker, R. R. (1992). Fraud detection: A theoretical foundation. *Accounting Review*, 753–782.

Patcha, A., & Park, J.-M. (2007). An overview of anomaly detection techniques: Existing solutions and latest technological trends. *Computer Networks, 51*(12), 3448–3470.

R Core Team. (2014). *R: A language and environment for statistical computing*. Vienna: R Foundation for Statistical Computing.

Riani, M., Atkinson, A. C., & Cerioli, A. (2009). Finding an unknown number of multivariate outliers. *Journal of the Royal Statistical Society: Series B (Statistical Methodology), 71*(2), 447–466.

Rosseeuw, P. J., & Van Zomeren, B. C. (1990). Unmasking multivariate outliers and leverate points. *Journal of the American Statistical Association, 85*, 633–639.

Rousseeuw, P. J. (1985). Multivariate estimation with high breakdown point. *Mathematical Statistics and Applications, 8*, 283–297.

Spink, J., & Moyer, D. C. (2011). Defining the public health threat of food fraud. *Journal of Food Science, 76*(9), R157–R163.

Tennyson, S. (2008). Moral, social, and economic dimensions of insurance claims fraud. *Social Research*, 1181–1204.

Wang, C., Viswanathan, K., Choudur, L., Talwar, V., Satterfield, W., & Schwan, K. (2011). Statistical techniques for online anomaly detection in data centers. In *12th IFIP/IEEE International Symposium on Integrated Network Management (IM 2011) and Workshops* (pp. 385–392). IEEE.

Wilks, T. J., & Zimbelman, M. F. (2004). Using game theory and strategic reasoning concepts to prevent and detect fraud. *Accounting Horizons, 18*(3), 173–184.

Index

© Springer Nature Switzerland AG 2019

F. Cecconi, M. Campennì (eds.), *Information and Communication Technologies (ICT) in Economic Modeling*, Computational Social Sciences,
https://doi.org/10.1007/978-3-030-22605-3

D

MIX
Papier aus verantwortungsvollen Quellen
Paper from responsible sources
FSC® C105338

If you have any concerns about our products,
you can contact us on
ProductSafety@springernature.com

In case Publisher is established outside the EU,
the EU authorized representative is:
Springer Nature Customer Service Center GmbH
Europaplatz 3, 69115 Heidelberg, Germany

Printed by Libri Plureos GmbH
in Hamburg, Germany